Geographische Zeitschrift, Special issue 2004

Editorial

On the occasion of the IGU Congress 2004 in Glasgow, the editors and publishers have assembled a selection of German-language articles on human geography which have appeared in the Geographische Zeitschrift in recent years. To make them available to a wider public they have been translated into English. The Geographische Zeitschrift publishes the results of theory-led empirical research from the broad spectrum of human geography. It is also an important forum for ongoing theoretical and methodological debates in human geography. Most articles are published in German, because of the growing significance of language and understanding for conducting differentiated discourses in the context of the "spatial", "cultural" and "linguistic turn". Hence, German geographers' resultant need to use their native language. There is, however, a growing trend to publish contributions by foreign authors, usually in English. Thus the Heidelberg *Hettner Lectures* are published regularly. The Geographische Zeitschrift has a special connection to this lecture series, as the journal was founded in 1895 by Alfred Hettner.

In selecting the contributions for this volume, the guiding principle was of variety as far as theoretical grounding, the objects of research and branches of human geography were concerned. It reflects the self-perception of the Geographische Zeitschrift, which wishes to document the liveliness, topicality and usefulness of human geography through its openness to new topics, and its scientific screening of contributions in accordance with international practice.

This special issue of the Geographische Zeitschrift would not have been possible without inspiration and help from many sides. The editors wish to thank the IGU German National Committee for providing the impetus for this issue, the German Research Foundation for its financial support, and the publishers for their willingness to take a business risk. Particular thanks are due to the authors: they placed their trust in the Geographische Zeitschrift by offering their research, and at the same time made clear their conviction that they would find a suitable platform for their ideas here. Our thanks also to the translators, whose excellent work was essential to the presentation of these contributions to a wider public.

Eike W. Schamp (Executive Editor),
Hans Gebhardt, Detlef Müller-Mahn, Jürgen Pohl, Ulrike Sailer, Dietrich Soyez

Geographische Zeitschrift, Special issue 2004

Space and Compatibility

HELMUT KLÜTER, Greifswald

Summary

The projection of economic and social structures onto space and territories (representing surveyed space) must be seen as a technically complicated and financially costly operation. The output of spatial abstraction includes

- the spatial orientation of other social systems,
- the co-ordination of activities within large organisations, the storage of information in a framework of space-time co-ordinates, and
- the compatibility within society.

Compatibility among formal organisations can be supported by binding property – together with mobile structures (such as persons) – to immobile pieces of land. The acceptance of territorial boundaries as limitations to unusual, risky activities, the communication, the modes of co-ordination and the final institution of rules for compatibility in society determine the patterns of spatial structures and their reflections as regions.

Thus, global society is not only characterised by global flows of information, finance, and trade, but also by a universal acceptance of spatial boundaries limiting all of these things. Spatial boundaries guarantee the compatibility of social systems with contradictory structures or contents. Spatial compatibility allows the privatisation of technical, informational and organisational advantages. As the programming of formal organisations grows increasingly complex, the process of spatial planning has turned from linear objective to functional planning, and then to compatibility oriented planning. Compatibility oriented planning is considered successful even if a problem is not solved but delayed or transferred to another region (the "anti-sustainability axiom" of modern spatial planning). This kind of planning stabilises differences between centre and periphery and other forms of stratification in society. Thus, compatibility oriented planning can be regarded as a survival strategy for public planning, especially in agglomerations, but not as an alternative to spatial planning that would guarantee equal living conditions and the development of democracy in space.

1. Introduction: space as a framework for society – society as a framework for space

At first glance there appears to be little connection between space and compatibility. The one term is derived from mathematics, architecture or geography, while the other is from materials science, biology or programming. This distinction by discipline suggests serious conceptual differences between the two quantities. But this view is unjust. If we approach space in its simplest form, we always encounter Felix Hausdorff's separation axiom, which requires that for every two distinct points x and y there must be a neighbourhood U of x and a neighbourhood V of y that are disjoint (Hausdorff 1898, pp. 60–62). Hence x and y are separate, but they fit together under certain conditions. Consequently it follows that there is compatibility with regard to the distinct points. Nevertheless, this simple, quite amicable sounding basic concept appears to contradict many geographical and sociological studies. In geography the fields of "lebensraum" and geopolitics defend ethnographically or politically founded incompatibilities as ontological. The focus of interest is not on the functionalised spatial boundary, but on the absolute boundary, which lays claim to validity for all sectors of society. The political boundary is intended to be simultaneously an economic, cultural and religious boundary, and one that is appropriately controlled.

In sociology, space in the most varied forms, from milieu to large city, is the precondition and justification for more or less insoluble sociological questions (cf. Stichweh 1998). Space is thus subjected to even less scrutiny than in traditional geography. The ontology that is evidently accepted by both sides proves practical from the point of view of scientific theory. Geographers have a great deal of practice meanwhile in relegating what is evil or inappropriate to sociology. With increasing globalisation and/or idyllisation, sociology is discovering geography for itself in a structurally rather similar manner. In the process, the category space is being used as a general container for framework conditions and thus serves as a refuse dump for what sociology finds inconvenient. We will not go into the crosswise trash function of the two disciplines in more detail here. Instead we will show how the category space can be derived with the aid of other concepts and axioms, namely in the following steps:

- Social loading of physical facts with spatial abstractions
- Spatial abstractions as programming aids for formal organisations
- Territorial bonds as a mechanism for creating and maintaining social compatibility
- Anti-sustainability axiom of regional planning.

The attempted derivation of the concept space from organisational theory also has consequences for the understanding of regionalisation and region. The traditional region concept ("region is something medium-sized that has developed as a region")[1] and the even simpler concept of a region as the diminutive form of society are contrasted with a regional concept that is functionalised via societal communication.

2. Abstraction of physical facts

If "space" is not supposed to be an ontology, an externally legitimated phenomenon, we must ask who introduces spatial categories into the discourse when and for what purpose. Implicitly Niklas Luhmann's systems theory offers a non-ontic way to theorise about space, for in his list of the dimensions of cognition the spatial dimension is lacking. For Luhmann the factual, temporal and social dimensions suffice. When he relinquishes space as a dimension of cognition he confirms the much older findings of the protocybernetician and sociologist Alexander Bogdanov-Malinovsky (1873–1928) and the already mentioned mathematician Felix Hausdorff (1868–1942). Since 1898, at the latest, the concept of absolute space is metascientifically no longer acceptable. At least in the work of these authors space is not an ontic category, but rather a derived one. Figure 1 is a systematic portrayal of the purpose, import and context of spatial argumentation.

Hausdorff's combinatorial mathematics (the factual dimension) is complemented by the temporal and social dimension (cf. Klüter 1986). In the process space is treated as a particular information strategy, whose special feature consists initially in the fact that the combined points are thought of as synchronous, i.e. simultaneous. This simultaneity, however, abstracts from the possible mobility of the points. The abstraction consists practically of stopped time. Naturally one can connect several spatial abstractions in a chronological series. This does not, however, in any way change the fact that the individual abstractions consist of synchronous structures. The duration of the synchroneity itself is relative. In geology such a time line can comprise several millennia; in plasma physics perhaps only a fraction of a second. Here the context of the abstraction plays a role – and it, for its

1 This concept of region exists meanwhile in a great variety of variants. Cf. Stiens 2000. For misgivings about it: Hard 2002.

	Structure formation	*Dimension of cognition*	*Theoretical reference*	*Disaggregation of of the concept space into*
Combination	.. of at least 2 points with neighbourhoods that do not completely overlap (modelling as a juxta-position of points)	Factual	Here: theory of social systems	– "Kulisse" (interaction) – Programme space (organisation) – Linguistic or code area (society) *
Synchronisation	... of those combined structures on a time line (= modelling as simul-taneity, as events parallel in time; immobility model)	Temporal	Evolution theory	– Topologies – Metric networks – Geometrical areas
Addressing	... of those synchronised structures for control pur-poses (spatial orientation, e.g. for regulating mobility)	Social	Communication theory	– Administrative space (politics; *Might, Right*) – Land, complementary region, anonymous address field (economy; *money, property*) – Landscape (*art*) – Home, small town idyll, ecological idyll (family, local association; *trust, love*) – Fatherland (people, school, armed forces, sport; *faith*) **

*(Types of social systems in parentheses), **(Sub-systems of society and *codes* in parentheses)

Fig. 1: Space as a special information strategy

part, can be narrowed down according to the needs of the person addressed. A geologist's understanding of synchronisation differs from that of a plasma physicist; that of a textile dealer bedevilled by seasonal ups and downs from that of the operator of a power station, who calculates in amortisation periods of decades.

The differences in information and intention ("sense") between the speaker and the system or person addressed is an important criterion determining which information is phrased in which terminology. The selection of the combined points out of all possible points is also subject to such prestructured criteria of meaningful and meaningless communication. Hypothetically it can be assumed that spatial orientation is one of the goals of such communication. The main goal is some kind of social orientation of the addressees, with the motive, e.g., of instigating a certain action. With the help of spatial orientation this motive is made anonymous, or the reason behind it is enormously simplified. This will be explained in the following.

The designation of certain time lines (e.g. the duration of validity of certain laws) implies differentiating between points that do not move during the period of observation and points that do move. The mobile ones are generally excluded or transformed into immobile ones, for instance as vectors or mobility areas. Because people are particularly mobile, they quite frequently fall outside of the area portrayed in spatial abstractions. Already this superficial impersonality suggests objectivity – although the selection and designation can often be quite subjective. To load physiogeographical facts with semantics is not a procedure that is limited to human geography. Physiogeography must also name the objects it deals with – and that is usually a social process. In our context it is significant that in so doing, scientific facts with their explicitness and objectivity are projected onto earthly space (territory). This objectivity shows up today in a great number of highly technicalised data and indicators, the most important of which are the metric scales for length, height and width. Since the standardisation of the metre and earth's co-ordinate systems, at the latest, the thus measured areas are no longer doubted socially. Their claim to objectivity is practically equivalent to the laws of physics. From this point of view it is understandable that many social systems – above all formal organisations like public authorities and companies – attempt to intensify their claims to validity and jurisdiction by means of territorial bonds (cf. Fig. 2). A river is then not only a river, but also a national boundary. On the multihued map of Europe its course determines the boundary of a state that claims legitimacy and validity. The state insinuates with this projection that it is just as objectively existent as the river.

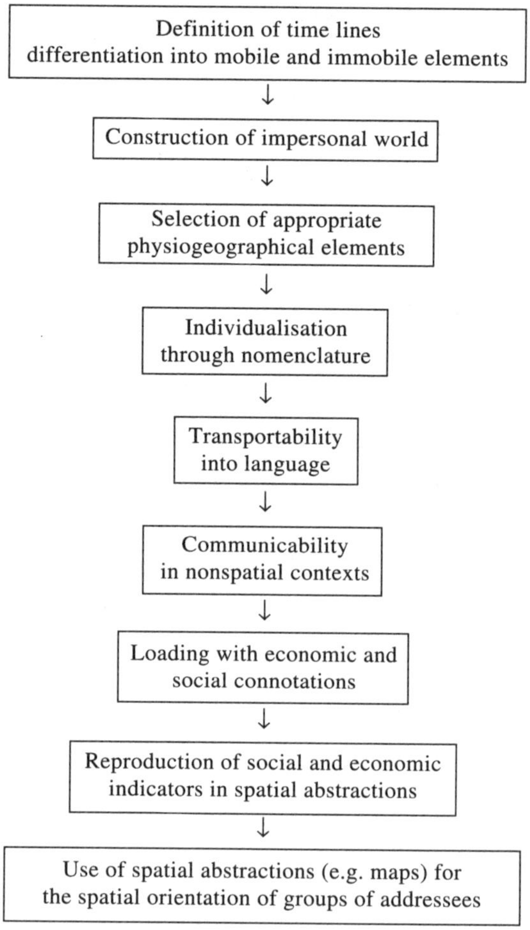

Fig. 2: Creating spatial orientation through selective spatial abstraction

As an example: Hamburg is both a fairly large settlement starting at its place-name sign and a large red dot on the map – though in everyday speech we do not attempt to differentiate between these two fundamentally different things. Since the one Hamburg indicates the other Hamburg quasi-empirically, a person reading a map will assume that Novosibirsk, which is also shown in a map, similarly exists, though he has no previous experience of the place. The direct connection between the person doing the recognising and his earthly surroundings (subjective spatial experience) thus covers only a small portion of his orientation substrate.

The empirically postulated knowledge of earthly regions that we nevertheless justify in this manner contains enormous inductive inferences from the sceneries that we have already experienced and their cartographic modelling to a world we have never seen. This presupposes unquestioning trust in scientific or other media organisation.

3. *From "Kulisse" to "programme space" (Programmraum)*

What I have termed "Kulissen" (literally scenery or backdrop, as in a theatre; used here in a figurative sense; cf. Klüter 1986, p. 54–59, 171) are the simplest type of spatial orientation in social systems (cf. Fig. 3). They serve primarily to structure the environment of interaction systems. Interaction systems are composed of at least two persons communicating with one another. To give an example, we have a person who asks the way to a certain place. The answer he receives is "first street on the left, then the third house after the streetlight on the right side". This may be sufficient for him to find the way. Our person does not need any geometrically ordered environmental information. More simply structured topologies are sufficient, and these are what we shall refer to as "Kulissen". The purpose of spatial orientation in this example is to help our person find a specific technical establishment (house), where something is to be done. Spatial orientation is not the main purpose of the action; it is only a means to realise a goal that may be quite "unspatial". Because of its incidental nature, the spatial orientation may not be consciously registered by the interaction system. This also means, however, that it is highly functionalised; it is employed for a purpose and is usually easy to decode – like the scenes of a play at the theatre. Hence spatial orientation is not an end in itself but serves to influence social objects (people, groups, companies, organisations) by means of certain information strategies (e.g. maps). That a map and the physical environment coincide is ensured not only by changing the scale during decoding. More important appear to be markers, by means of which socially relevant control semantics are artificially placed in the physical environment. Only certain organisations (territorial authorities, especially road construction authorities) are allowed to set up and maintain such markers. Our capacity to perceive the physical environment soon runs up against boundaries, despite a variety of generalisation strategies. Markers can be placed in such a manner, however, as to allow us to incorporate into our repertoire of activities even processes that are actually biologically impossible – e.g. the high speeds that are technically possible on autobahns. The letters on the signs are so large that they can be made out even at 200 km per hour.

Markers and technology stimulate the self-control capacity of the persons to whom they

Types of social systems	***Selection according to***	***Stabilisation by***	***Time range***	***Minimum requirement for spatial orientation***	***Structure of spatial orientation***
Interaction	Presence	Persons, themes	Very short-term	**"Kulisse"**	Organisational relations via certain environmental elements
Organisation	Membership	– Positions – Rules of membership, – Basis of decisions (programme)	Development: – Past: history, tradition – Present: decision – Future: planning	**Programme space**	Topological or geometrical reproduction of certain environmental elements; exactly fitted to addressees
Society	Communicative accessibility	Today: functional differentiation into new subsystems	Evolution	**Linguistic or code area**	Areas of spectrums of validity, comprehensibility and use

Fig. 3: Spatial orientation in certain types of social systems (space as a combinatorial problem)

are addressed. The always existent external control mechanisms (regulatory objective of the creator of a spatial abstraction) can be blurred organisationally or by means of information, or made almost completely anonymous. As an example we can take the driver of a car who gets his bearings with the aid of a road map and does not notice the external control input until he is stuck in a traffic jam on the autobahn. He allowed himself to be seduced into taking the autobahn by the thick line on the map without even considering alternative routes on smaller roads. Who is at fault? The cartographer? The traffic conditions? The driver (because he didn't take the train)?

The answer is of necessity very complex, because the road system and the traffic regulations are so designed that if something occurs, the injured party or victim often becomes aware of self-regulation deficits on his part or of personalised external ones. The accident insurance system is also so devised that it attempts to determine the individual responsibility of the involved drivers. How and why the autobahn perhaps has only four lanes at this place and not the six lanes that would help to avoid accidents is certainly the last question asked. And even when it is asked, the question is again that of the individual responsibility of construction engineers and planners.

The reconstruction of our "case" neglects systemic relations. From the many variants of behaviour conforming to the rules one cannot infer how the system functions correctly. Even the opposite can be the case: if all drivers in the above mentioned traffic segment had exceeded the speed limit with a certain regularity perhaps neither a traffic jam nor an accident would have occurred. Spatial abstractions and the spatial orientation based on them can thus totally obscure causal relationships. This is true not only in traffic, but also within large companies and agencies. Complicated production processes are divided up into space/time co-ordinates for the individual labourer and employees. Their activities do not require complicated definitions of purpose, means or goals. Instead they can be directly defined via the co-ordinates. For instance, within the period from 7 AM to 3 PM 5000 metal pieces have to be bent on punch y. Punch y can be explicitly identified in a ground plan of the plant. Explicitness implies that the labourer in question will not go to punch z, which is next to punch y. Punch z is intended for another labourer. Both have to have explicit, i.e. objectively comprehensible, spatial perceptions. This spatial abstraction is drawn up by the plant management when it programmes all labour processes and must be decipherable for all employees. Explicitness can, for example, be ensured by geometric scales in the spatial abstraction. Such abstractions of "programme space" replace a large number of individual commands and thus minimise management's control efforts (cf. Klüter 1999). Not least because of the above described impersonality, but also because of the pseudo-objective relations to the physical environment via empirically comprehensible markers, spatial orientation often functions without hierarchisation. It works as an externally defined necessity composed of only few real and many "virtual" obstacles. For the management of organisations it is a typical soft power instrument. These advantages are the reason why in modern societies spatial abstractions are employed, or newly developed, for many control tasks, despite the great effort required to develop them. Their production and reception are extremely asymmetric. Whereas objectifiable abstractions of physical and social facts always require an organisational input in the conception, production, drawing and reproduction of space, spatial orientation can occur completely without organisation in interaction systems or even personal systems. From this is derived the organisational a priori concept of the production of socially effective spatial abstractions. The organisational benefit of spatial abstractions far exceeds the programming and control of subordinate or associated systems. The thus produced synchronisation patterns are used for controlling, because here once more the problem of capacity created by the simultaneities arises. For large organisations socially dimensioned time vectors are not sufficient because of the many, usually communication dependent, overlaps. Consequently controlling usually falls back on a time-place co-ordinate grid. This again predestines "programme space" structures to serve as a storage medium. Wolfgang Hartke's idea of the landscape as a photographic plate on which the social geographer leaves traces (Hartke 1959) takes on a new variant here: the "programme space" becomes a hard disk, on which the formal organisation stores its past.

4. Compatibility as equivalent of causality and functionality?

In our above mentioned example of a traffic jam and the accidents it caused, the manner with which the matter is dealt with by the law or by insurance companies shows how difficult it is to divide highly organised processes into cause-effect relations. Evidently the only events that can be decided are those that can be attributed to concrete deeds performed by persons. This means, theoretically spoken, that formal organisation must be reduced to interaction systems – which is by definition not completely possible. Particularly problematic are the "virtual" constraints created by the rules. In the example we took from traffic, the driver of the car in back could perhaps have avoided a rear-end collision if he had dared to swerve onto the hard shoulder, crossing over the solid boundary line on the right. In this case the illicit alternative would have been quite reasonable, because all motorists are familiar with the programming of the road traffic system. The programmes of smaller firms, for instance the time targets in the trucking business, are usually a trade secret. For the employed drivers the schedule is nevertheless a quasi law. The lorry driver in our example perhaps drove too fast in fog – quite aware that his accident risk is covered institutionally by the company, as far as this is possible. The rules of one system contradict the rules of the other system, etc. The causalities that can be derived from this likewise contradict each other. Many of them are not obvious from outside because the company's operational programmes are kept secret. The mere quantity of such processes already forced us to switch from causality to functionality as permissible control concept. Morally-legally it is perhaps reprehensible to ruin one's fellow-man. If, however, in the sense of maximising profits this occurs between a bank, which suddenly recalls a credit, and a borrower who cannot repay, it is permissible.[2] Here the assumption that all are striving to increase their material prosperity helps. The differentiation of society into sectors, such as economy, politics, religion, art and science, makes allowances for the varying functionalisations, leads to different programming requirements and thus to different spatial abstractions (cf. Klüter 1986, pp. 109–166).

When no ordered conditions can be created, even via functionalisation, the only solution is separation. Separation can even ensure the compatibility of conflicting systems. Compatibility itself can evidently be the last maxim of functionalisation. Guaranteeing compatibility emerges as the last alternative to mutual destruction. On the other hand, compatibility provokes more complicated forms of social co-operation, for instance mutual functionalisations as to how to stabilise non-co-operation, i.e. how to draw the demarcation line. Compatibility of itself is unproductive. An example can explain this. Until early modern times most of what people did and the people themselves were considered to be fully functionalised. For certain purposes almost anything could be done with them: they could be forced to work, tortured, sold, possibly killed, married off, etc. Not until the rights of

[2] Cf. Shakespeare: The Merchant of Venice.

Mechanism of compatibility	***System-environment gradient[1] based on***
Inaccessibility	Physical, social or technical distances or barriers prevent collisions
Legal obligations	Human rights, Civil law, corporate law, especially contracts, Commercial law Constitutional law, international law, etc.
Memberships, social inclusion/exclusion	Associations, societies, corporations, certain types of enterprises, functional exclusion by means of the appropriate systems
Territorial boundaries	Very elaborate form of exclusion with many prerequisites, the definition of the boundary being the result of complex social practice
Special arrangements for border crossing	Bold communication (e.g. journalism, science), criminality, security maintenance, state of emergency, financing

[1] Social systems have no rigid boundaries. Their range of social and technical communication decreases continuously, not abruptly. Limiting factors are the attention or interests of others.

Fig. 4: Creating compatibility among social systems

man were formulated and implemented was a person per se granted the right of existence. Merely because he is a person he has a right to exist. In religion, philosophy and law this new convention created a vast number of problems. The result was that a person as an intrinsic value was ultimately inaccessible to any kind of direct functionalisation. Making individual people compatible with each other became a task of the legal system. The legal system became the functional equivalent of the primeval forest, the desert, paradise, the vast distances, in which mere inaccessibility ensured the compatibility of individual people (cf. fig. 4).

With the emergence of the law of contract and of corporate law this construction was also applied to formal organisation. Since Max Weber and the constitution of the Weimar Republic, at the latest, associations, societies, political parties and other organisations have enjoyed the same sovereign rights as persons were granted through the rights of man. A political party no longer defines itself through its programme, but through the fact that it is a party. Even the will to exercise political power is not absolutely requisite. An association can call itself "homeland association" even if it destroys the economic basis of the homeland – for instance by exaggerated hostility to foreigners, with the result that no young labourers move into a region inhabited by a high percentage of elderly persons. The society can define away problems with foreigners for itself – for instance by not admitting foreigners. This type of creation of compatibility via membership or prohibited membership is something like the conditions that were customary before the rights of man were introduced, for the sovereign rights of the organisation imply the lack of freedom of its members. In First Division football we might even say that slave trading has been revived to a certain extent, considering the procedures by which players are transferred between football clubs. Otherwise, teams are formed; they fight against each other in a sporting manner, create rules of fairness, which rediscover one human right or another – but only as a game, not by any means in the work of the team off the playing field.

If all of these compatibility mechanisms fail, one can create the paradisaical inaccessibility artificially by erecting a fence or a wall. This functions in cases in which the territorial bonds of the parties to the conflict were previously institutionalised. In many societies the institution of human rights was closely linked to the institution of private ownership of land. In the Netherlands, France, Germany, Russia and South America this mixture was the fuel for revolutions. Having territory of one's own allows the owner to do on his piece of the world what is prohibited elsewhere. An employer buys obedience from his labourers. When they cross the boundary of the factory, they automatically change from free citizens into more or less zealous production factors. The territorial boundary is one of many system boundaries. The advantages of such a regulation seem to be so great for the owners that they accept the limitations to their own mobility. This in turn induces the state administration to buy loyalty from its citizens by allocating property as broadly as possible – for instance via programmes encouraging the construction of private homes. The greater the amount of private property, the greater is the need for protection by the administration. Property disciplines. One could for example assume that extreme right-wing views are hardly less widespread in some rural areas of western Germany than in eastern Germany. In the west, however, such views are not allowed to spread beyond the pub. Here people are generally owners of farms, which in the sense of "property" and "inheritance", are used for credit creation. If a person were caught in a right-wing activity and punished for it, this would threaten his credit standing at the bank and with it everything "he has achieved". In eastern Germany neither property ownership nor the resulting disciplinary effect is currently so widespread.

5. Planning and standardisation of programme space

Using compatibility as our guiding theme allows us to move away from the system as a model of equilibrium. One can imagine compatibility even between extremely unequal components that cause, at least temporarily, extreme lopsidedness in the system as a whole. For example, the communication medium right/might appears capable of stabilising even very drastically unequal distributions in the medium money/property. One could almost imagine that large parts of the legal system were created expressly for this purpose. Precisely at this point territoriality and compatibility play a special role. The difference between those who are owners of landed property and those who are not was and is socially a problem zone whose composition varies greatly historically, nationally and locally. In formal organisations planning is therefore usually doubly entrenched, for one thing in the medium might/right and for the other in the medium money/property. For the sector politics plots of land are the smallest unit of administrative space; for the sector economy they are projections of monetary value onto geometric areas (cf. Klüter 1987). The cadastral system and tax laws define the area of overlap. Already in 1978 Luhmann pointed out the limits to functional argumentation in the organised planning process (Luhmann 1981, pp. 367–369). On one hand this reflects the fundamental problem of the division of labour: at the absolute present we cannot know how others have decided at this moment and to what extent older causalities (experience) and functionalisations (limiting contingency via objective functions) were realised. We work with more or less probable assumptions, more or less solid structural linkages, more or less promising calculations and with increasing shifts into the more reliable technological, mechanical, automatic spheres. Precisely the latter is particularly relevant for planning, because more refined technology raises the speed of monitoring. To be sure, in the process the present does not become more compact, but the past becomes more rapidly recallable. This in turn allows more contemporary work and stronger delegation into the future. We make compromises in the present to the detriment of the future. An example might be a case in which a raise in the tobacco tax is postponed by three months at the request of the tobacco industry. With this it may end up in an election year and thus be automatically abandoned.

This creation of temporal compatibility through planning can be done analogously for the spatial sphere. One person makes compromises in his own programme space at the expense of others. An interesting example is Germany's decision to phase out nuclear power. The deadlines for shutting down the atomic power plants may lead to shortages of electric power. This does not mean that new safe conventional power plants are being built. Instead, once the power market had been liberalised, the power-supply companies looked around in other countries. Inevitably they discovered Russian atomic power plants. Particularly lucrative contracts are those with the operators of power plants of the Chernobyl type. The plants are generally not as well maintained as the actual Chernobyl power plant, but they are inexpensive. Sosnovy Bor near St. Petersburg is probably being revitalised for this very reason. This means that to ensure the future power supply in Germany, relatively safe German nuclear power plants are being shut down and in exchange not so safe Russian ones are being reactivated. One can only hope that when it comes to the next nuclear accident, the wind won't be blowing from the Northeast.

The externalisation of unpopular decisions is, however, only one of the merits of the creation of spatial compatibility. Another one consists in the establishment of buffer zones. Completely contradictory activities can be compatible with each other, for instance the loud, smelly, dirty production of plastic objects and cosy life in a detached house, if a park or a small wood is inserted between them to serve as a green finger. A motorway with noise barriers can also fulfil this function, or a former waste dump. Undesired, negative environmental influences need not necessarily be externalised. They can be neutralised by appropriate encirclement. The public planner thus not only plans a given activity, but he plans how to make this activity compatible with other already established activities. Precisely this is what modern land use planning boils down to. The less it can deal with substantive or goal oriented plans, the more it plans compatibilities. In this manner the creation of spatial and temporal compatibility adds up to an anti-sustainability axiom of modern regional planning. One subsists not only by solving problems but also by relegating problems to the future or into programme spaces that are not one's own or by encircling them so as to isolate them.

According to this logic, world society cannot be exclusively reduced to the fact that everything or many things are accessible to communication. It also exists because that which is theoretically achievable is not really achieved and remains unachievable. Despite globalisation somebody has to fortify the river Oder so that the imaginary floods of immigrants from eastern Europe will not inundate poor Germany.

Here it becomes apparent that the system-environment gradient of organisations can be stabilised by boundaries, but that it cannot be reduced to them. Of the spatial references of a business enterprise only the plot of land can be reproduced with clear geometric demarcations. Its complementary region consists formally of networks of subcontractor and marketing relations that far exceed the plant's plot of land; its information or address field is composed of topologically ordered lists (cf. Klüter 1987) that encompass a much larger area. It is not actually a problem that the areas of the complementary and information field of one company overlap with those of others, because these are subject-oriented projections and relationships that are not bound by legal titles to specific territories. Only if the capacity of the infrastructure that they need to use together for these relationships is limited is renewed organisational input required. The stabilisation of a group of organisations cannot therefore be conceived directly as a matter of working together or against each other – e.g. in the sense of competition. A common interest must exist. Practically, this usually means that an organisation manages to establish its

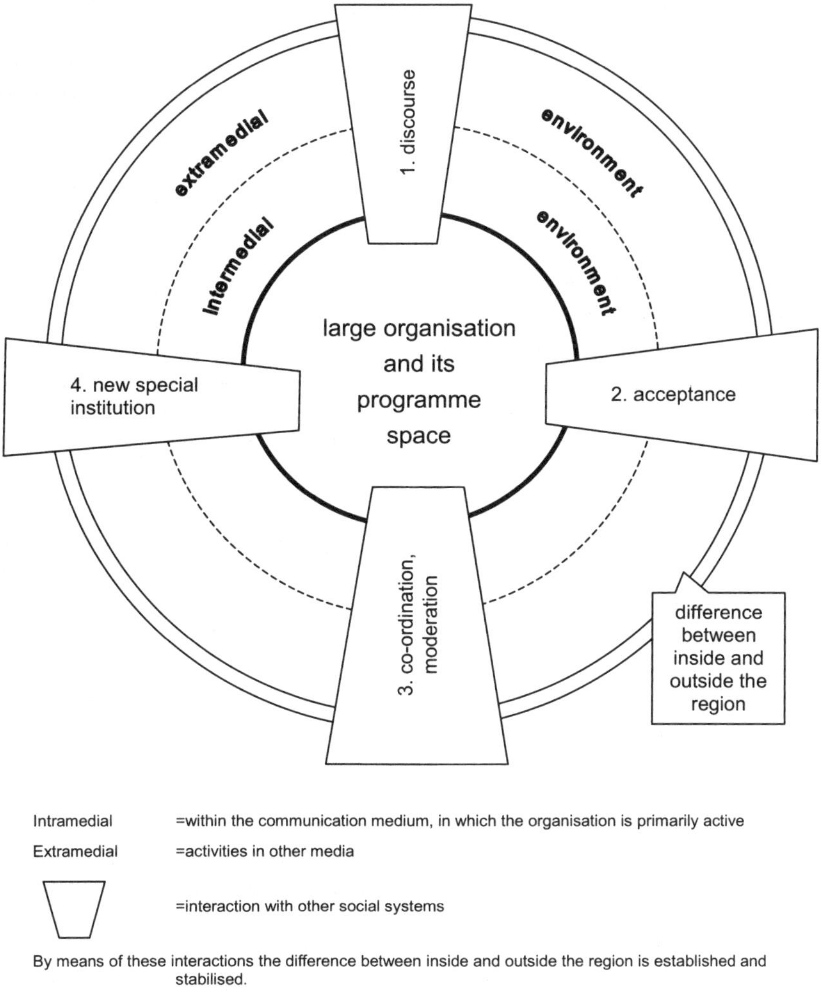

Fig. 5: Making a region (clock model)

programme space and its surroundings as the standard and acquires a monopoly of universally valid signs in physical space (cf. Fig. 5). To be able to do this the organisation must have adequate weight and must have existed in its milieu for an above average length of time. It is thus large organisations that avail themselves of such a role, particularly providers of infrastructure or businesses that are particularly active in communications, such as newspaper publishers, television companies, etc. By means of discourse they first standardise their intramedial environment, which affords them the necessary acceptance. Such media of communication (e.g. codes) are money (in the sector economics), might/right (politics), aesthetics (art) and others. Only when the boundaries of a programme space of one enterprise or organisation have been accepted by other enterprises can they adapt themselves to this. With this reputation and the associated boundary system one can influence other organisations that use other communication media. Thus certain co-ordination and moderation services are possible between organisations that belong to different social sectors. For such services a new independent organisation can be established. It will attempt to stabilise itself intra- and extramedially, i.e. to assert itself in other sectors of society with other codes. On the whole the organisational milieu that evolves develops more secure internal communication that protects it from external influences. In case of a new impulse from outside or an innovation from inside, the course of discourse, acceptance, moderation and possible further institutionalisation is set in motion again. If it proceeds along the old, already institutionalised lines, the region is stabilised. Its outside demarcation is intensified. A region is thus the programme space (or an element of the programme space) of a formal organisation (e.g. a territorial authority, a state agency, a newspaper publisher or another business) that has become a standard in social communication and is used as frame of reference for the activities and plans of a large number of other organisations. The region is thus no longer "society in miniature" or a reservoir of social processes, but instead a strategy for stabilising the system-environment gradient of a large organisation in continuous communication with others.

Once their standardised co-ordinates, markers and territorial boundaries have been "physicalised" – i.e. become as objective as physical geography – (cf. Fig. 2), many of their programme points can be portrayed as inherent necessity, as given facts or simply physical constraints. As a rule they are accepted in this form by other members of society without further contemplation and precisely for that reason they can realise their control capacity.

In this connection the political nature of the co-ordination and moderation process applying above the level of the individual organisation need not necessarily be a natural given (cf. Luhmann 1997). The institutionalisation as *polis* derives from the western city model. Closely related to this, but with a different historical motivation, is the concept taken from agrarian societies of including pieces of land in the administrative space and the resulting regional and/or state structure. Where neither continuous agrarian nor urban territories were available, other large organisations served as norm.

- Thus, for instance, in ancient Prussia it was claimed that the army kept a state (as its financial base).
- In Russia the decisive advance took place after 1917 when, in accordance with the GOELRO plan for electrification, the power plant companies, i.e. the infrastructure providers, improved and optimally organised the energy supply. They defined the territory of the early USSR as a set of energy supply regions.
- With their work for flood prevention, infrastructure firms in the Netherlands played an important role in the development of the national polity.

Today almost all modern states are not only political moderators but complex technical, social and economic infrastructure concerns that substantially influence the economy and other sectors of society (cf. Simonis 1977). Their infrastructure technology and communication ranges seem to determine the territorial size of regions.

The region as an element of the programme space exclusively of large state organisations is not completely convincing. By means of various territorial boundaries individual state agencies can carve up the administrative space in such a manner as to make it difficult for regions to form. This is, for instance, the case in the eastern part of the German federal state of Mecklenburg-Vorpommern. Here planning regions and county boundaries disregard historical boundaries. The districts of the pension offices have different boundaries from those of the state building authorities and these in turn from the police districts and they from the environmental protection agencies, the road construction agencies, etc. (cf. Landesregierung Mecklenburg-Vorpommern 2003, p. 23 ff). The intention behind these measures, which were pursued in the early 1990s, was to integrate the Vorpommern portion of the state into the Mecklenburg part. For this reason no clear political regionalisation of Vorpommern was to be permitted. Enormous co-ordination deficits and moderation losses were implicitly accepted, for which the state government in far away capital Schwerin was unable to compensate. The result is a situation that sounds paradoxical. Although it had the better starting factors, for the first time in recent history since around 1997 the population of Vorpommern has had a lower per capita income than that of Mecklenburg. The disparities between the two parts of the state are growing rapidly. Currently the unemployment rate in Western Pomerania is 10 percentage points higher than in the western part of Mecklenburg. The regional product per inhabitant in Vorpommern is 57% of the national average.

According to the traditional geographical concept of a region, the state government acted correctly. To prevent regional consciousness and regional identity it did not permit large institutions to use the term "Vorpommern" in their name. What was not foreseeable with the traditional approach was that the deficient co-ordination and control in the eastern part of the state would lead to a previously unknown degree of corruption and to parallel and counteractive subsidisation. Not least this finding forces us to take Benno Werlen's warning seriously that we should not only concern ourselves with experiencing and interpreting regions but also with the question for what and why they were made (Werlen 2000). The new government of Mecklenburg-Vorpommern has therefore formulated as one of its reform goals "Einräumigkeit", i.e. the creation of the same programme spaces for different state administration authorities (Landesregierung Mecklenburg-Vorpommern 2003, p. 14).

6. Some results

Neither geography or mathematics nor the theory of science can provide an abstract concept of a given space. On the contrary, spaces are structured as the environment of specific social, technical, psychological, personal or biological systems (cf. Klüter 2003).

Most practical spatial concepts employ an abstraction of physical factors, which even in their simplest forms are subject to social or technical selection.

Spatial abstraction (encoding) and spatial orientation (decoding) are extremely asymmetrical. The encoding of socially relevant spatial abstractions must be organised in companies and public authorities. Spatial abstraction creates a programme space as an important element of self-organisation, because by this means it is possible to programme extremely effective connections between external control and self-control of other systems and persons addressed.

With the spatial abstractions they have received, the addressees plan their own movements. This spatial orientation complements, simplifies or substitutes for many forms of social orientation. The high likelihood of success makes spatial abstractions an important controlling instrument for formal organisations.

This in turn is the basis for the fact that many large organisations also use their programme space for saving data and creating their past – as a "hard disk" as it were.

Because of the continuously higher speed of processing and access to programme space storage, in many cases planning periods can be shortened. Things that were planned long term ten years ago are planned for the medium or even short term today.

As the programming strategies are refined, the possibilities for postponing things to the future or shifting them to other programme spaces increase exponentially. This "anti-sustainability axiom of regional planning" makes functional, content oriented planning difficult.

Under the complex present-day conditions the great number of social systems, especially formal organisations, necessitates special strategies for synchronisation (e.g. division of labour) and for creation of compatibility. Compatibility channels the risk of conflicts and creates security.

A compatibility strategy that requires a lot of effort but is easy to handle, legally and addressee oriented, is to link social systems to territoriality. Bonds to territory simplify the abstraction of human mobility. The coexistence of segmentally ordered plots of lands implies simplicity and mutual compatibility in an otherwise enigmatic society. Besides, it is simply and clearly conveyed. The drawing of territorial boundaries itself is highly functionalised socially and technically.

The externalisation and/or isolation of problematic masses can therefore likewise be dealt with by means of territoriality. The functional equivalent of a problem solution is to expand the refuse dump or the ghetto.

Thus territoriality as a compatibility strategy is one of the basic instruments for securing and privatising organisational advantages in world society.

The search for absolute security arouses an interest in absolute territorial boundaries that are to be valid for all sectors. From the point of view of a functional systems theory this is nonsense. This demand is repeatedly raised in ideologised contexts ("geopolitics"). Not least here it becomes obvious how problematic an ontological space concept or "space as society in a microcosm" is, when it is made into a programme as an end in itself.

Limits to capacity in the formation of social networks and the basic assumptions of technological milieus in which organisations always also act (infrastructure) encourage the development of regions as an element of communication in the system-environment gradient of large organisations. A region is thus the programme space (or an element of the programme space) of a formal organisation (e.g. a territorial authority, a state agency or a business) that became a norm in social communication and is used as the reference framework for the activities and plans of a large number of other organisations.

The need for co-ordination can be concentrated in new regional institutions that perform monitoring, controlling and storage functions and concentrate activities externally for both political and economic enterprises and associations.

A region allows planning to be restricted to the conditions and needs required for compatibility in the co-ordination or management of differing interests. The success of planning does not require an exact cause-effect analysis via internal system regulation and development processes.

For personal and interaction systems spatial abstractions are preferentially orientation strategies. For formal organisations they are primarily programming, controlling and information storage strategies. On the level of society they serve as essential mechanisms for creating code-specific compatibilities, "code" meaning here the languages of component systems of society, money, might, right, art, trust, etc. (after Luhmann). With this the idealistic attainability algorithms of a "world society" are evaded. The territorial bonds of social systems and compatibility rules about territorial boundaries are very effective mechanisms with which concrete organisational advantages and other extreme system-environment gradients can be maintained over long periods of time.

The system environment, composed of many addresses, private claims, real and virtual obstacles, becomes, in the form of organisation-specific objectified space, a ghost train of society in which persons, action systems and organisations are imperceptibly pressed on to the rails of societal feasibility and compatibility.

References

Bogdanov, A. A. (i.e., A. A. Malinovskij) (1925): Vseobshchaya organizatsionnaya nauka – tektologiya, 3. ed.. Leningrad: Kniga.

Hard, G. (2002): Landschaft und Raum. Aufsätze zur Theorie der Geographie Bd. 1. Osnabrück.

Hartke, W. (1959): Gedanken über die Bestimmung von Räumen gleichen sozialgeographischen Verhaltens. Erdkunde 13, pp. 426–436.

Hausdorff, F. (1898): Das Chaos in kosmischer Auslese. Erschienen unter dem Pseudonym Paul Mongré. Leipzig.

Klüter, H. (1986): Raum als Element sozialer Kommunikation. – Giessener Geographische Schriften 60. Giessen.

Klüter, H. (1987): Wirtschaft und Raum. G. Bahrenberg et al. (eds.): Geographie des Menschen – Dietrich Bartels zum Gedenken. – Bremer Beiträge zur Geographie und Raumplanung 11. Bremen, pp. 241–259.

Klüter, H. (1999): Raum und Organisation. Meusburger, P. (ed.): Handlungsorientierte Sozialgeographie. Benno Werlens Entwurf in kritischer Diskussion. Stuttgart, pp.187–212.

Klüter, H. (2003): Raum als Umgebung. Meusburger, P; Schwan, T. (eds.): Humanökologie. Ansätze zur Überwindung der Natur-Kultur-Dichotomie. Stuttgart, pp. 217–238.

Landesregierung Mecklenburg-Vorpommern (ed.) (2003): Verwaltungsstrukturreform M-V: In Zukunft einfach besser – Eckpunkte zur Reform der öffentlichen Verwaltung im Land Mecklenburg-Vorpommern. Dokumentation des Beschlusses der Landesregierung vom 21. Januar 2003. Schwerin.

Luhmann, N. (1981): Organisation und Entscheidung. Vortrag. Erneut abgedruckt. N. Luhmann: Soziologische Aufklärung 3. Opladen, pp. 335–414.

Luhmann, N. (1997): Die Gesellschaft der Gesellschaft. Bd. 1–2. Frankfurt a.M..

Simonis, E. U. (ed.) (1977): Infrastruktur. Köln.

Stichweh, R. (1998): Raum, Region und Stadt in der Systemtheorie. – Arbeitspapiere 30 der ZWE "Arbeit und Region", Universität Bremen. Bremen.

Stiens, G. (Hg.) (2000): Die neue Konjunktur von Region und Regionalisierung. Informationen zur Raumentwicklung, Heft 9/10.

Werlen, B. (2000): Alltägliche Regionalisierungen unter räumlich-zeitlich entankerten Lebensbedingungen. Informationen zur Raumentwicklung Nr. 9/10, pp. 611–622.

Address of the author: Prof. Helmut Klueter, Department of Geography, University of Greifswald, F.-L. Jahn-Str. 16, D-17487 Greifswald, Germany

First pulished in Geographische Zeitschrift vol. 90/ 2002 (3 + 4) pp. 142–156.

Geographische Zeitschrift, Special issue 2004

Entrepreneurship research
The relevance of the region and tasks facing economic geography

ROLF STERNBERG, Cologne

Summary

Start-ups are one of the most relevant current subjects in broad areas of science, economics and, in particular, politics. There are many reasons for this which may be found in such areas as new technological developments and their economic implications as well as the general structural shift and the increased significance, despite the globalisation trend, of very small firms. This paper analyses the arguments for the proposition that the determinants behind start-up activities are not independent of spatial factors. On the contrary, spatial factors can be identified, systematised and – in principle – quantified among both the causes behind and the economic impacts of start-ups. Economic geography in particular is therefore confronted with many numerous research topics in the fields of theory, empirical research and policy. Economic geography has comparative strengths over disciplines also concerned with start-ups of which it should take advantage in a more proactive and self-confident manner than has been the case to date.

1. Start-ups and young firms as a current topic in science, the economy and politics

Start-ups and young firms have recently become a focal point in politics, economics and science for various reasons. Politics only discovered start-ups fairly recently, but then quickly developed a range of initiatives and programmes. Whereas start-ups are taken for granted as a practical institutionalisation of the dream of independence held by immigrants to the USA, European countries such as the United Kingdom have only recently placed the subject of entrepreneurship at the centre of their economic policies, as demonstrated by the Blair government's Competitiveness White Paper (published by the Department of Trade and Industry in 1998). The strategic consequences of the weaknesses identified for the United Kingdom – too few of the admittedly high number of start-ups grow quickly and too few of those start-ups which do grow quickly originate from large companies or universities – are also currently popular among most of the German political parties. As a logical consequence of this, a successful entrepreneur such as Jost Stollmann, the founder of Compunet, was being openly discussed as a possibility for the post of Minister of Economic Affairs until shortly before the last elections to the Bundestag (German lower house) and, as an equally logical consequence, the subject of entrepreneurship was made an election campaign topic which did more damage than good to the incumbent (traditionally more business-friendly) government. Hopes in Germany are that the numerous promotional programmes at national, Land (state) and municipal level will make a positive contribution to the development of the labour market. Examples of these programmes include the federal government's "Exist" programme and the "ERP" equity aid programme at national level, the "GO" start-up drive in North-Rhine/Westphalia at the state level (the institute for research of small and medium-sized firms – Institut für Mittelstandsforschung – offers an initial evaluation of this programme for the year 2000) and, at municipal level, the start-up centres in approximately 150–200 municipalities around Germany (see Sternberg et al. 1997). Start-ups became a burning topic in politics partly out of conviction (ambitions to create an "entrepreneurial society") and partly out of necessity (the realisation that large companies in the past have made job cuts, while start-ups really can only grow). The university entrepreneurship chairs established with the aid of public funds are also an indication of the increase in political awareness of start-ups.

Finally, business has recently (once more) become increasingly interested in start-ups. The growing significance and / or introduction of such phenomena as the New Market, business angels, risk capital, the new economy or business incubators as private versions of the familiar public start-up centres are all evidence of this development.

It is impossible not to notice the increased activities in the field of start-up research in a broad range of scientific disciplines, as is demonstrated by the growth in the number of publications, journals and congresses on the subject in recent years. There is also evidence of this trend in the field of research into economic geography within Germany and abroad (see Sternberg 2000b). The German Research Foundation (Deutsche Forschungsgemeinschaft – DFG) responded to this development with the establishment of a Priority Programme "Interdisciplinary Entrepreneurship Research", for example. This may be interpreted as a reaction to the many years of ignorance in broad areas of all disciplines regarding start-ups' status as a special segment within the business sector and the resulting gaps in research on the subject.

This paper analyses the relationship between the (economic) region and start-up activities from a theoretical and empirical point of view and from the angle of economic geography. The following core chapter initially addresses fundamental cause-and-effect relationships between start-up activities and spatial aspects, after which empirical evidence, predominantly from German regions and studies, is discussed. The last chapter draws conclusions for economic geographic research, with the emphasis on new areas of research.

2. Entrepreneurial activities and spatial factors – discussion of an interrelationship

2.1 Fundamental cause and effect relationships

Three subject areas dominate current research of entrepreneurship (see also Preisendörfer 1996):

- the description and explanation of start-up activities (how much and why),
- the description and explanation of the success / failure of start-ups (intensity and causes of the development of the firm),
- the conceptual structure of and monitoring the success of policies to promote entrepreneurship).

As can be seen in overview in figure 1, there is currently a range of theories and approaches which attempt to explain start-up activities and their success (or otherwise). The analysis of start-up activities and the propensity to launch new firms was for a long time dominated by approaches concentrating on the supply side and personal factors relating to the entrepreneur, with the emphasis in particular on the motivation and motives for launching a company. These factors only stood up to empirical analysis to a limited extent, which is one of the reasons why demand and environmental factors (including regional factors) have gained in significance in recent times. The trinity of the character and behaviour of the entrepreneur, the characteristics of the start-up itself and the nature of the environment in which the start-up is launched is currently broadly accepted when considering the factors of success (cf. Klandt 1984) and is standard in empirical analyses (cf. most recently

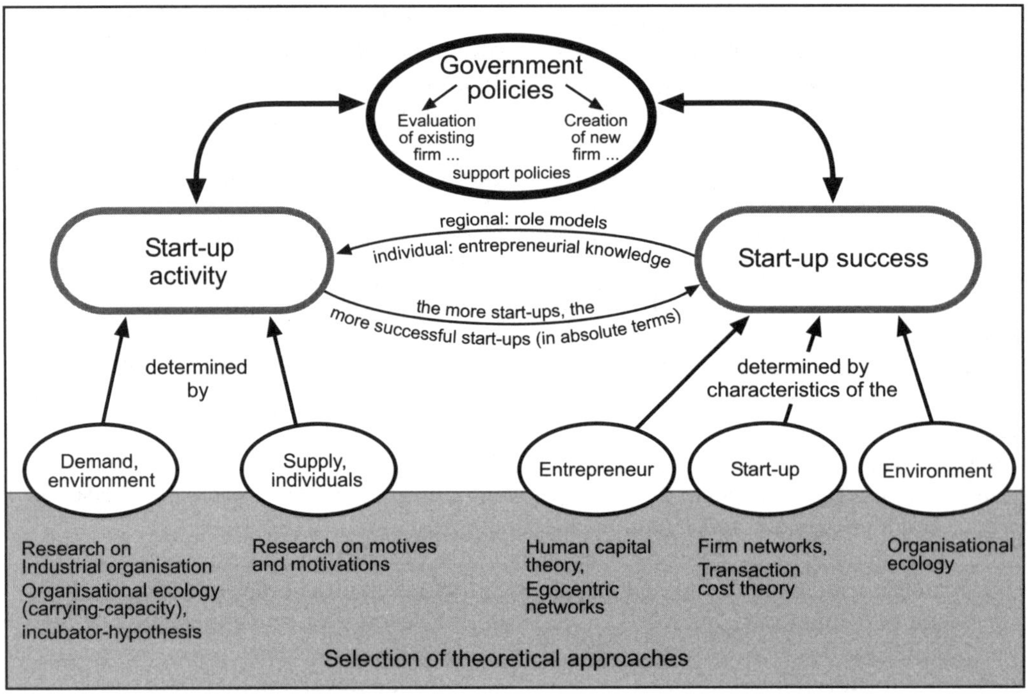

Fig.1: Issues and theories in entrepreneurship research. Source: Preisendörfer 1996, significantly modified

Schutjens/Wever 2000). It should be emphasised that the determinants of the success of a start-up and start-up activities (or propensity, or probability) differ considerably, an aspect which is overlooked by not a few, even empirical, studies. For each of these factors there are concurrent attempts at explanation or partial theories which do not explicitly, or only rarely implicitly-, take into consideration the issue of the spatial causes or impacts of start-ups (for example as an element within the environmental factors). This is no surprise in view of the dominance in entrepreneurial research of micro and macroeconomists and – some considerable distance behind – sociologists and psychologists, with economic geographers tending to date to be "on the peripherals, making various individual contributions" (Preisendörfer 1996, p. 9).

An apparent paradox may be observed from the point of view of economic geography in the current era of globalisation (cf. Schamp 2000): The increase in the significance of highly flexible, small and often recently launched, knowledge-intensive firms is accompanied by a drastic increase in the number of large multinational companies, caused in part by an unprecedented wave of mergers. But these two developments need not necessarily be contradictory in nature. Schumpeter's view of an innovative entrepreneur, the concept of the long waves of economic development, which can also be interpreted regionally, and the natural uncertainties in times of a technological and economic paradigm shift make it possible for the coexistence of very small firms (a category which includes most start-ups) and large companies to seem plausible. Specifically focussing this observation on the relation between the region and start-up activities has still yet to be done theoretically. There is still no generally accepted theoretical approach to the explanation of start-up activities at both the national and the regional level (cf. Wennekers/Thurik 1999 for the national level). For a long time, researchers of entrepreneurship coming from the economic sciences, for example, paid too little attention to the spatial factor and, vice versa, economic geographers ignored the subject of start-ups. The current situation, particularly with regard to regional policies to promote start-ups, is the result of this trend, for there is a considerable and unsatisfied demand for regionally differentiated, empirically supported studies concerning the target groups of these policies. Most empirical studies on entrepreneurship to date refer to the national level as a standard scale, ignoring the interregional differences within a particular country. One can hope, however, that times are changing in this respect: the issue of spatial proximity ("Geographical proximity" following the French school led by Torre/Gilly 2000) has become very modern in particularly those areas of economics whose protagonists consider themselves to be a part of the "new economic geography" (see Karlsson/Dahlberg 2003). This is indicative of the fact that the economic sciences no longer explain the performance of companies solely or primarily on the basis of determinants related to the firm itself or to the entrepreneur. The consideration of environmental factors in a broad sense, including spatial proximity and features of the regional environment, is becoming evermore prevalent and popular. According to this increasingly influential school of thought, entrepreneurship is a "generically social, a collective phenomenon" (Johannisson 2000, p. 306) that cannot be explained solely through the attributes of individual persons. Rather, entrepreneurship must be explained with recourse to the social capital concept and the actor networks related to it, which create positive externalities. This is an area in which economic geography still has a great deal to contribute to the relationship between entrepreneurship and regional development processes (cf. Bolton/Westlund 2000).

The points put forward in this chapter are intended to support this theory and are based on the following argumentation. All other factors remaining equal, the propensity of a person in a given region to launch a new firm is higher, the greater the number and size of incubator institutions there are offering sufficient fertility in that particular region and from which it is possible to launch a firm. Existing start-ups, however, also benefit in their development from a positive regional environment, which includes as preconditions not only incubator institutions (availability and attitude towards start-ups), but also an appropriate climate for start-ups. As part of a self-perpetuating process (including factors such as the role model effect of successful start-ups and their intraregional networking), regional clusters of start-ups may start to develop in which, due to agglomeration economies and other positive external effects associated with spatial proximity, start-ups develop better economically than outside the clusters. With a view to national competitiveness, it would therefore be ideal if start-ups were to concentrate within a controlled number of regional start-up clusters. The number of these clusters depends certainly on the existing demographic structure, which only changes over the long term but, viewed over the course of history, can be expected to remain relatively stable. The make-up of these start-up clusters may change over time, much in the same way as a lifecycle which applies not only to products, but also to economic regions. These days, for example, Munich is indisputably the number one high-tech region in Germany; 100 years ago Munich had no claim whatsoever to this title. Conversely, once dynamic growth regions with high numbers of start-ups have become areas of old industry with only very low levels of entrepreneurial dynamism.

Fig. 2 demonstrates that there is an interdependency between entrepreneurial activities within a region on the one hand and economic development, or the characteristics of this region, on the other.

The upper section of this diagram shows potential effects which start-up activities may have on the regional economy (for example, birth and mortality rates with the correspond-

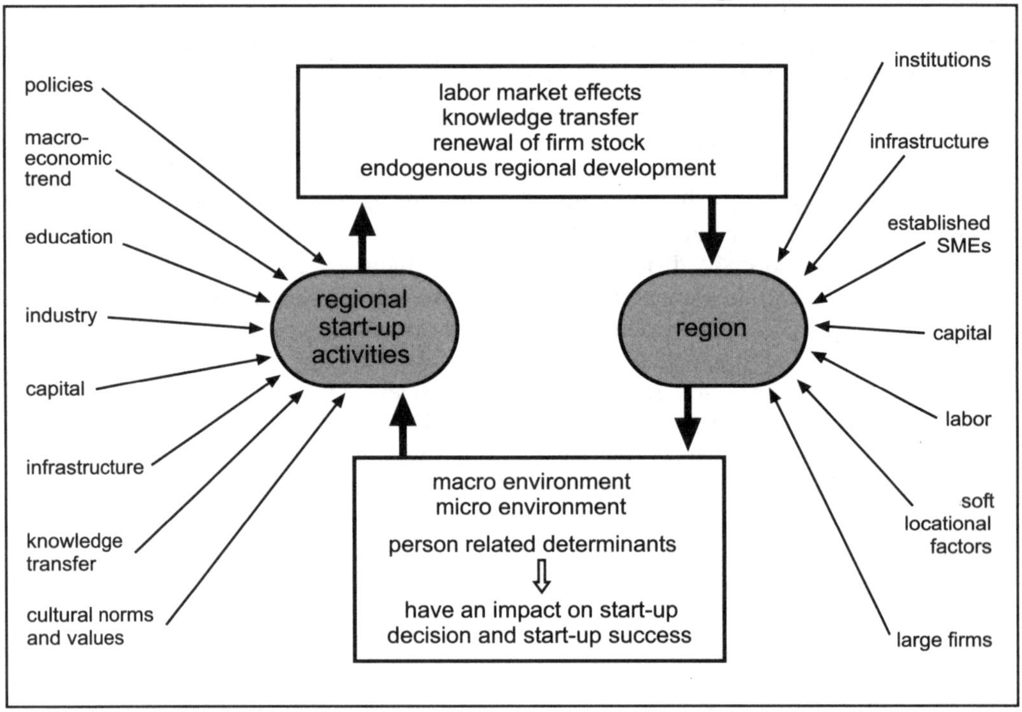

Fig. 2: Causes and effects of start-up activities in regions

ing employment rates, as well as the growth and decline of young, existing firms). Effects on the labour market are predominantly generated – according to the widely held opinion which has prevailed since the thesis of David Birch at the latest – by the total of small firms, which explicitly includes start-ups. Young firms in particular generate employment initially in the regional environment where they were established, due to their high level of regional embeddedness. Knowledge transfer between knowledge producers / knowledge-producing institutions (e.g. universities) and knowledge users is at its most efficient between individual heads – transfer between individual people. And it is most creative if the person, for example from a university, undertakes self-employed activities in his or her own firm, instead of entering dependent employment, and puts his or her own ideas into practice. The target group here is therefore knowledge-intensive start-ups. Start-ups of this nature contributed to the evolution of entire high-tech regions such as Silicon Valley or the Greater Boston region, which would not have grown to such an extent if it had not been for the start-ups which originated from Stanford University and the Massachusetts Institute of Technology (MIT) and established themselves locally. The renewal of the stock of firms emphasizes a different aspect. Schumpeter's concept of the creative destruction of traditional economic structures by dynamic entrepreneurs – a concept which originally related to entire macroeconomies – can, in principle, be applied to individual regions. Finally, a reference to the subject of "endogenous regional development" should be made. It is easy to overlook the fact that there is a strong endogenous component to regional development concentrated on start-ups, i.e. very young firms. Based on the spatial immobility of start-ups, which has been clearly proven empirically, they actually form part of the potentially "established stock" of the region and are therefore the specific object of efforts to retain them. This kind of endogenous regional development allows a certain amount of independence from developments outside the region (e.g. the decisions taken by corporate headquarters far away from the regions), but this "established stock" has to be the target of constant efforts to retain it and, even if these efforts are successful, the result is nowhere near as spectacular as if a large company decides to move to the region.

There are, of course, many other factors which have an impact on regional development, some of which are the product of established theories regarding regional economic growth and some of which, however, are more recent. Each of these growth determinants listed in the right section of figure 2 consists of a component within the region and a component outside the region. Capital may originate in the region, for example, but it may also come from outside the region. In addition, the same external influence does not generate the same effect on regional development in every region. Analysis of the relative weight of the influence of start-up activities on regional development as compared with that of the other intraregional or extraregional determinants would be a worthwhile task for regional scientific research, particularly as it would probably produce different results from one region to another.

The lower and left sections of figure 2 address the question of how the regional environment, in other words the characteristics of the region, can influence regional start-up activities. There are many factors which influence the propensity (of an individual and as a total of all individuals in the region) to launch a new firm and the success of the start-up. They are systematised as follows in literature on the subject:

- macro-social environmental factors: culture, society, politics, scientific infrastructure, economic structure;
- micro-social environment: social and professional origins, networks;
- personally specific factors: motivation (push vs. pull factors), demography (age, gender), aspects of personality (willingness to work hard, take risks).

As a matter of principle, these factors have an influence in all countries and in all regions of a country. They have a clear regional dimension, however, because they come into play to varying degrees and thus operate differently in different regions. An example will serve to illustrate each of the three groups of factors. A demographic feature as an example of person-oriented factors is the distribution of age and gender. The features vary between countries as well as between sub-national regions. The social and professional background of potential entrepreneurs as an example of microsocial environmental factors is also partially regionally dependent, as every region has a specific social and professional structure. The high degree of spatial immobility of entrepreneurs is correctly interpreted as evidence for the importance of the regional environment for start-ups (cf. Lageman et al. 1999).

Regional influence is most obvious in the macrosocial environment. The degree to which sufficiently large and thematically related incubator organizations (e.g. universities, innovative firms) exist in the regions varies greatly and should therefore be expected to differ in its effect on the propensity to establish a new firm from one region to another in the same way as, for example, the difference in market proximity differs in its effect on the success of start-ups in individual regions. Prevezer (1998) shows impressively that for regions in the USA, the quantity and even more so the quality of academic R&D institutions (in the study, the number of scientists in the research institutions in the Life Sciences area that exhibit content-based affinities to biotechnology) in the places investigated were the most important factors for explaining the market entry of biotech start-ups. The local emergence of biotechnology can be unequivocally traced back to leading researchers at certain research institutions. The proximity to these "key star scientists" was the most important reason behind the market entry rates of the late 1970s. This can be shown by looking at the two leading regions in biotechnology at that time, the San Francisco Bay area (including Silicon Valley with Stanford University and UC Berkeley) and the region of Boston, Massachusetts. The high costs of these regions, already high-tech regions with a plethora of typical agglomeration diseconomies, notwithstanding, settlement was profitable due to synergy effects. Empirical evidence for the high relevance of the macrosocial regional environment can also be found for Germany and for less technology-intensive industries, as Moßig, for example (also published in this issue) identifies local spin-offs as the origin of clusters of the German packaging machine industry. Most of these original spin-offs do not settle just anywhere – exceptions of interregional exports of knowledge confirm this rule – but within a tight spatial area around their incubator institution.

In general, the opinion of Frick/Lageman/von Rosenbladt/Voelzkow/Welter (1998, pp. 80) is accurate, that "intermediary institutions" as part of the regional infrastructure provide an entrepreneur with the necessary services which he needs when setting up a firm, but does not himself possess: "They are generally based in the regional environment of the firm and, either directly or indirectly via contacts, offer functional or resource-specific support for selected problems" (ebd., p. 80). The intermediary institutions operate via such actors as public and semi-public promotional institutions (chambers, associations, technology and start-up centres, transfer institutions, research institutions), private organisations (investors and providers of debt capital such as banks) and politics, and have a positive or negative influence on the establishment and development of start-ups. In spite of all the debates about the political promotion of start-ups through institutions such as start-up centres or universities, it should not be forgotten that the most important incubators in terms of quality for rapidly growing start-ups are still other firms and that – in strong correlation with this – people with experience of launching start-ups do so particularly successfully (in other words, the start-ups grow quickly), as demonstrated recently by Rosa/Scott (1999) in the case of Scottish "local heroes".

The characteristics of a region can in principle influence both the propensity to launch a new firm and the economic success of endogenous start-ups. One of the most important contributing factors to the growth of many of today's high-tech regions, particularly in the USA (Silicon Valley, Greater Boston), but also in Munich, to take a German example, is the strong entrepreneurial dynamic compared with other regions in the same country and the same time (see Sternberg 1998 for an international comparison of such high-tech regions). An important reason for the considerable significance of entrepreneurs for the development of such regions is the spatial immobility of such spin-off start-ups, which is mainly attributable to the personal ties of the entrepreneurs to "their" region and is the cause of many regional high-tech clusters, as recently demonstrated by Hendry/Brown/Defillippi (2000), for example, for six optoelectronics clusters from three countries. Regions with comparatively high rates of entrepreneurship in particular demonstrate that not only favourable overall economic framework conditions are responsible for these high rates – these conditions also benefit other regions in the same country – but also, to a considerable extent, the regional entrepreneurial and innovational climate (see Lessat et al. 1999), the integration of new firms in the regional environment (see Lompe et al. 1998) and the infrastructural characteristics of the region. Intraregional network relationships play an important role in this respect for young firms and their founders: specific (often also regionally specific) knowledge, frequently decisive for the development of the start-up, is generated in these networks, as recently demonstrated empirically by Collinson (2000), for example, with the example of Scottish software start-ups from the multimedia sector. It is certain that not only intraregional environmental factors influence entrepreneurial activities. It would be unwise to support a regional determinism. As is the case with regional development in general, monocausal explanations are out of place here. Here, too, there is a lack of empirically representative research that presents a detailed analysis with respect to the entrepreneurial activities in a region of the relative importance of intraregional factors on the one hand and supraregional factors on the other.

2.2 Empirical evidence: Start-up activities in Germany's regions

Regional entrepreneurship research plays a inappropriately small role within the field of entrepreneurship research in Germany and has to date been concerned primarily with the quantatively very small, but politically particularly interesting segment of innovative start-ups, or even technological start-ups and their regional distribution, as well as regional causes of start-ups and their effects (cf. for example Nerlinger 1998).

The regional distribution of the absolute number of start-ups is first and foremost a representation of the size of the region in question (population level, number of people in gainful employment), which is why large metropolitan areas in particular are dominant. In order to counteract this effect of scale and to allow interregional comparisons of the intensity of start-up activities, it is helpful to use an additional reference value which puts the analysis into proportion. The most common methods are the "business ecology approach" (taking the number of businesses as a reference value) and the "labour market approach" (taking the number of gainfully employed people as a reference value). Figure 3 shows the intensity of start-up activities calculated on the basis of the number of people suitable for gainful employment, presented by regional planning regions. The selection of this reference value is intended to attach adequate significance to the fact that the unemployed (the number of whom is the main element in the difference between those suitable for gainful employment and those actually in gainful employment) certainly should be counted among the potential entrepreneurs in a region. The existence of various political programmes (e.g. bridging money) demonstrates the empirical evidence of this entrepreneurial potential.

The picture for start-ups in technology-intensive sectors of the economy, which are of particular interest for politics, deviates in part from that for start-ups in general, both in absolute and relative terms. In an absolute view of the roughly 50,000 start-ups from the period 1994 to 1998, as contained in the data set of the Centre for European Economic Research (Zentrum für Europäische Wirtschaftsforschung – ZEW) based on the credit reference agency Creditreform (Foundation Panel for West Germany, Foundation Panel for East Germany), strong concentrations in the metropolitan regions of Berlin, Munich, Frankfurt am Main, Stuttgart, Rhine-Ruhr and Hamburg are obvious (cf. Sternberg 2000a). When looking at the intensity (cf. fig. 3), the spatial structure typical of many indicators of innovation can be seen. Within western Germany, there is an obvious North-South divide; for Germany as a whole, a West-East divide. The Munich regional planning region together with the region close to the German Alps is clearly in first place, as is generally confirmed by literature on high-tech industries (cf. Sternberg/Tamásy 1999).

Nerlinger (1998) showed, mainly on the basis of the same ZEW data set, that innovative entrepreneurs in western Germany are heavily concentrated spatially – more strongly than other start-ups. Innovative entrepreneurs prefer districts in or, increasingly, near densely populated areas where agglomeration advantages can be exploited without suffering the disadvantages of metropolitan areas. This statement tends not to apply to start-ups in general, i.e. not taking only those in technolo-

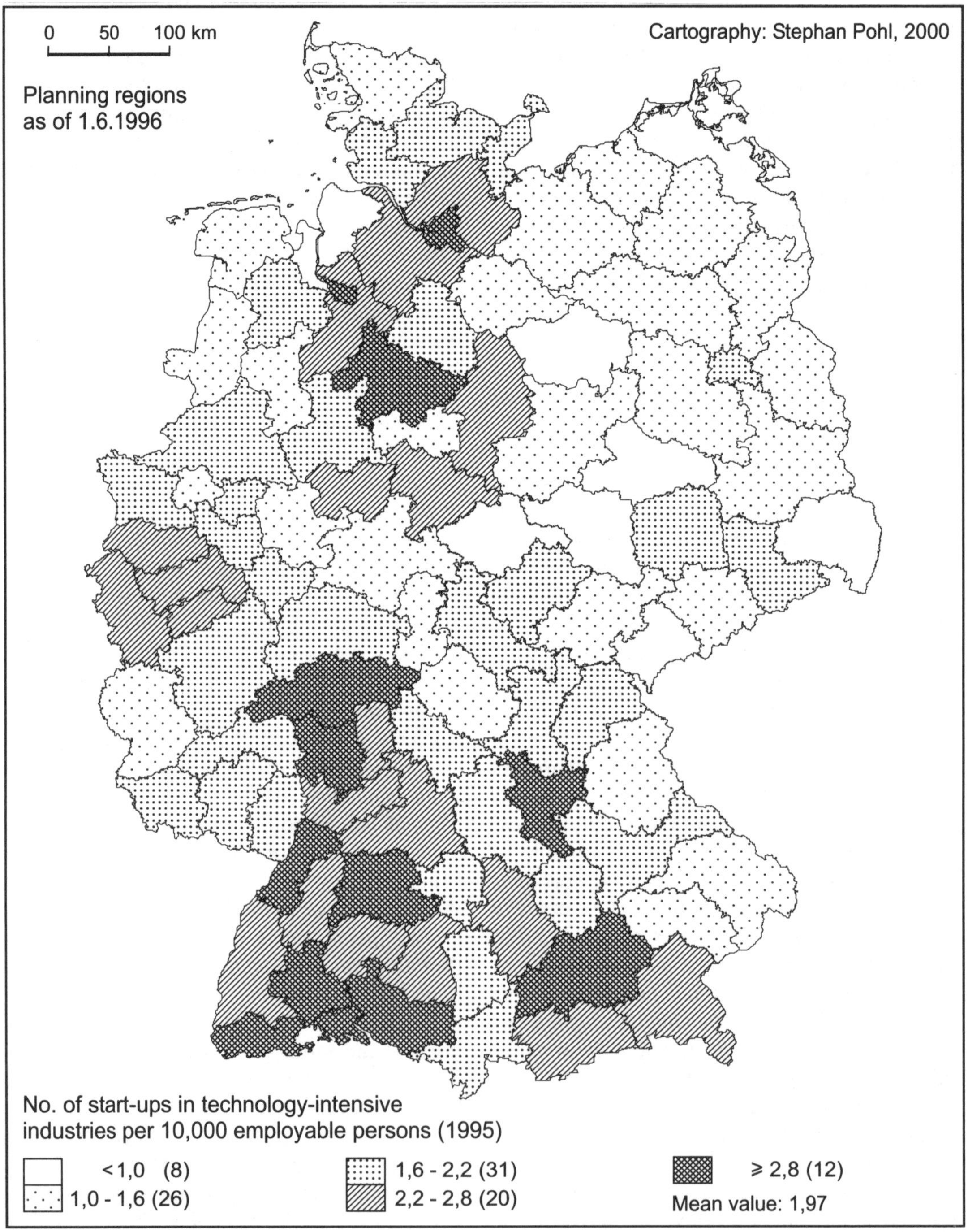

Fig. 3: Start-up intensities in German Planning Regions, 1994 to 1998 (technology-intensive industries)
Data sources: ZEW start-up panel West, ZEW firm panel East

gy-intensive sectors of the economy, as Fritsch (1997) demonstrates a clear centre-periphery divide for start-ups within western Germany. Fritsch/Niese (also published in this issue) demonstrate using employment statistics that, depending on the choice of reference value, either agglomerations (using the "business ecology approach") or regional areas (using the "labour market approach") recorded higher entrepreneurial intensity in Germany between 1983 and 1997.

After this descriptive presentation of the regional distribution of start-ups in Germany, the following looks at the regional causes for this pattern. The fact that the regional environment can not only be a cause of regional differences in the frequency of entrepreneurial activities, but also tends to influence the interregionally varying effects of start-ups, need only be mentioned here for the sake of completeness. A good regional infrastructure for start-ups and a corresponding environment are factors in the success of existing young firms (cf. for example May-Strobl 1999). Regional characteristics can have a positive or negative influence not only on the frequency of entrepreneurial activities, but also on the "post-entry performance", in other words the development of the firm after launch.

In 1994, Reynolds/Storey/Westhead presented one of the most comprehensive internationally comparative studies of start-up ratios, taking into consideration regional causes. The results of this study show for the countries involved (France, Germany, Italy, Sweden, Great Britain and the USA) that start-up ratios between the individual regions of a country vary by a factor of between 2 and 4. There is only negligible difference between the causes of the regional variation in the start-up ratios in the individual countries. In Germany's regions, the start-up ratio has a negative correlation (!) with the regional development of employment, both based on the "business ecology approach" and with the "labour market approach", and a positive correlation with the proportion of small and medium-sized firms among the total number of firms (cf. Fritsch 1997 using the same data source).

Fritsch/Niese (also published in this issue) also attribute the differing regional frequency of entrepreneurial activities in western Germany between 1983 and 1997 to industry effects in particular as part of a shift-share analysis, as the start-up rates between industries differ considerably. The remaining complex of causes, interpreted as a locational effect, is considered relatively unimportant and shows rather unexpected results. The whole of Baden-Württemberg and the majority of Bavaria, for example (excluding Munich) achieve only relatively poor values for the propensity to launch a new firm when these values are interpreted based on the locational effect. Munich is an exception within Bavaria, however, and certainly the most dynamic of the all of the most significant entrepreneurial regions in Germany. Here it can be seen that the – predominantly not coincidental – interaction of various regional environmental factors has contributed to create a highly innovative milieu which has been beneficial not only to the regional economy as a whole (cf. Sternberg/Tamásy 1999), but also, explicitly, start-ups.

The suitability of macro-studies such as the works of Reynolds/Storey/Westhead (1994) or Audretsch/Fritsch (1994), as well as the shift-analysis method cited above, for identifying the causes of entrepreneurial activity in certain regions is very limited, however, as these causes are regionally specific, also with respect to individual industries and time. More fruitful in this respect are works analysing individual industries and / or regions. Consensus is spreading in broad areas of the regional sciences that regional entrepreneurial activities and innovation activities depend on the innovation regime of the industry in question *and* the region in question (cf. Lageman et al 1999). Zucker/Darby/Brewer (1994) demonstrate using the example of the young biotechnology industry in the USA, that both the location and the growth of the start-ups are predominantly attributable to locally available and – at least during this phase – immobile human capital. The few excellent biotechnology researchers at universities were the decisive factor in the USA between 1976 and 1989. Other factors, such as the local availability of venture capital or political programmes, the number of universities and those studying at them per se (i.e. regardless of the top researchers) do not go any significant way towards explaining start-ups in the biotechnology sector.

In this respect, it should be noted that industry structures and other environmental factors are not the only aspects which may be regionally specific, and therefore in a position to influence the entrepreneurial propensity in a given region, but that social and cultural standards may also be regionally specific. Saxenian (1996) emphasises this point in an impressive comparison of the two high-tech regions in the USA, Silicon Valley and Route 128 around Boston/Ma. She demonstrates that it is the specific west-coast mentality of entrepreneurs who have already failed once which makes it possible for them to gain risk capital from banks a second time, which is harder to do in Greater Boston. Even if this kind of interregional differences in mentality are less marked in Germany than in the USA, they certainly should not be ignored.

One of the subjects increasingly being analysed in Germany is the significance of knowledge spillover (cf. for example Anselin/Varga/Acs 1997 and, for western German regions, Funke/Niebuhr 2000), which, particularly in times of globalisation, is most effectively demonstrated within sub-national regions, such as at the city level (cf. Audretsch 2000). Applied to the subject of regional influences on entrepreneurial activities, this generates the question of whether it is not just innovative companies in general which benefit from the localised research capacities of other actors in the same region (companies or public research institutions), but specifically start-ups also. The correlation with the politically highly relevant and scientifically highly interesting subject of regional start-up concentrations or clusters is plain to see. At county level, there is a clear statistical correlation between the number of start-ups in R&D-intensive industries and the existence of teaching and research staff at universities and non-university research institutions – as a direct consequence of spin-off start-ups and spillover effects. Nevertheless, obeying the rules of regional labour distribution in Germany, start-ups in R&D-intensive industries (cf. fig. 3) are more broadly dispersed as a result of radiation effects than the R&D institutions (cf. Beise et al. 1998).

Regional competence centres in the sense of spatial/sectoral concentration of potential technological knowledge are increasingly being understood and promoted in federal research and technology policies, which could amplify spin-off effects (cf. Roland Berger & Partner 1998). When looking at the intensity of entrepreneurial activities in R&D-intensive sectors of the economy, a relatively high affinity can be seen in the regional patterns of start-ups in high-tech areas and in technologically intensive service sectors and in start-ups in sectors of the economy which are not R&D-intensive and those that belong to industries of advanced technologies. These pairs of industries are therefore in a complementary relationship and are advanced by the same factors (cf. Beise et al. 1998). The idea of competence centres in its explicitly regional version is suited for making clear the correlation between the economic effects of regional and national entrepreneurial activities. It should be emphasised, however, that use of the term "competence centre" in political and economic discussions is not currently unambiguously meant in the spatial sense, which has the same effect as undervaluing the significance of the region for regional *and* national innovation and entrepreneurial processes (cf. Sternberg 2000a, p. 196). As a result of various factors mentioned, both the launch of start-ups is more probable and the economic development of existing young firms is better within a start-up cluster than with a relatively dispersed distribution pattern. As an aggregate for all start-ups in all regions of a country, this means that start-up activities on a national scale (and their economic effects) are greater than with a relatively equal distribution over the whole country.

Statements made thus far in chapter 2.2 have been based on analyses of regional entrepreneurial activities, which in turn were based on secondary data. The following section introduces an ongoing research project which describes and explains (in some cases regional) entrepreneurial activities in international comparison based on the collation of primary data. Levels of entrepreneurial activities have an influence on national and regional growth processes, about the scope of which little is currently known. At the same time, the frequency of start-ups and their success is also influenced by the environmental factors mentioned in figure 2. These include a range of mainly non-economic determinants (often described as the entrepreneurial climate or in earlier times as the economic spirit). Both aspects – the contribution of the entrepreneurial sector to macroeconomic growth and framework conditions relating to start-ups – are the focal point of the Global Entrepreneurship Monitor (GEM), an internationally comparative research project running since 1999 and coordinated by the Babson College/USA and the London Business School. A Global Report is published once a year with an international comparison (most recently Reynolds/Hay/Camp 1999) and, shortly afterwards, the country-specific reports of the individual country teams. Empirically, GEM is based on comprehensive surveys of a representative proportion of the population (by telephone) and experts (in interviews), investigating entrepreneurial activities and attitudes as well as framework conditions relating to start-ups (financing, politics, etc.). Details on the methods and the results relating to Germany are available from Sternberg (2000a), who is also head of the German GEM team. Ten industrialised countries were involved in the GEM pilot project in 1999. In view of the positive response to the results in 1999, roughly 20 countries are expected to take part in 2000.

The data collated from the survey of the population in the pilot phase of GEM in 1999 did not allow any representative statements to be made about the regional differences in entrepreneurial propensity and attitudes in Germany. Far more telephone interviews were carried for the collation of data in 2000 in Germany than in 1999 and more than in other countries in 2000 (7,000 compared with 1,000 / 2,000). In addition, the German data set contains several regional variables (e.g. residential area of the respondent by districts), which allow regionally specific statements about entrepreneurial activities, the perception of the chances for start-ups and the image of entrepreneurs among the public.

The following analysis is based on the thesis according to which both the perception of start-up opportunities and the image of entrepreneurs may differ between regions and – partly for that reason – entrepreneurial activities differ from one region to another. This partially interdependent relationship between features of the regional environment, which is also expressed in the views of those members of the public surveyed, and entrepreneurial activities should then also be reflected in indicators of statistical correlations. Table 1 gives a summary of the results of the relevant tests.

Table 1 takes six variables into consideration which characterise the attitude towards start-ups and the entrepreneurial behaviour of those surveyed. In addition, the start-up ratio was calculated based on the three criteria for GEM, as was an index of the entrepreneurial climate. The latter combines the six exclusively dichotomous variables and defines them as follows:

$$I_i = ((0{,}007 \quad E_1) + (0{,}082 \quad E_2) + (0{,}093 \quad E_3) + (0{,}030 \quad E_4) + (0{,}054 \quad E_5) + (0{,}025 \quad E_6)) \quad 1000$$

where

i : regional planning region i where i= 1,2,3 ... 97

$E_{1...6}$: attitude variables in accordance with table 2

Tab. 1: Regional difference in entrepreneurial attitudes in Germany

Statements	Do statistically significant differences exist between the dependent variables of column 11 at the level of ...					
	the planning regions (97)?		the Federal states (16) ?		Eastern vs. Western Germany (2)?	
	Chi-Square[3]	significance[2]	Chi-Square[3]	significance[2]	Chi-quare[3]	significance[2]
Similar standard of living (V_1)	127,9	*	56,3	***	31,4	***
Fear of failure (V_2)	122,7	*	45,7	***	15,5	***
Good start-up opportunities (V_3)	306,7	***	181,4	***	135,6	***
Resentment against successful entrepreneurs (V_4)	125,8	*	25,8	*	16,2	***
Respect for entrepreneurs (V_5)	106,7		35,7	**	4,4	*
Business Angel (V_6)	84,3		18,4		0,0	
Entrepreneurial climate index[1]	2,6	***	8,5	***	83,5	***
Start-up rate	1,0		2,0	*	7,0	***

[1] The entrepreneurial climate index is the weighted sum of the assessments of the six statements, see section 2.2 for details of the calculation

[2] *: statistically significant at the 95 percent level of confidence, **: statistically significant at the 99 percent level of confidence, ***: statistically significant at the 99,9 percent level of confidence

[3] F-value of analysis of variance for entrepreneurial climate index and for start-up rate

Similar standard of living: In your country most people would prefer that everyone had a similar standard of living. (1= no, 0= yes)
Fear of failure: Fear of failure would prevent you from starting a business. (1= no, 0= yes)
Start-up opportunities: In the next six months there will be good opportunities for starting a business in the area where you live. (1= yes, 0 = no)
Resentment against successful entrepreneurs: People you know resent those who make a lot of money from starting a new business. (1= no, 0= yes)
Respect for entrepreneurs: People you know respect those starting a new business. (1= yes, 0 = no)
Business Angel: You have, in the past three years, personally provided funds for a new business started by someone else - this would not include buying publicly traded shares or mutual funds. (1= yes, 0 = no)

Data source: GEM population survey Germany June/July 2000 (N=7.034)

The contingence coefficients for the correlation between the entrepreneurial ratio (= number of subject persons who qualify as start-up entrepreneurs according to the GEM definition, 2.5 % in Germany in 2000) and the respective attitude variable are taken as weighting factors. As such, the fear of failure with one's own start-up and the individual's perception of the current chances for a start-up have the strongest and most significant correlation with the decision to launch a new firm. The higher the index value, the more favourable the entrepreneurial environment in the region.

Apart from one exception, all variables demonstrate in some cases highly significant correlations between the individual regional areas. Even at the level of regional planning regions, there are statistically significant differences in the perception of entrepreneurial opportunities and the summarial climate index, but also in the proportion of actual entrepreneurs. The perception of entrepreneurial opportunities is the one attitude variable where

Tab. 2: Entrepreneurial attitudes in Western and Eastern Germany

Statements[1]	Percentage of adult population with favourable entrepreneurial attitudes in Western and Eastern Germany[2]			
	Western Germany	Eastern Germany	Germany	(valid cases)
Similar standard of living (V_1): no	47.7	38.5	45.8	(5,871)
Fear of failure (V_2): no	57.4	51.0	56.1	(5,955)
Good start-up opportunities (V_3): yes	33.9	15.7	30.1	(5,195)
Resentment against successful entrepreneurs (V_4): no	86.4	81.7	85.5	(5,782)
Respect for entrepreneurs (V_5): yes	80.8	78.1	80.3	(5,858)
Business Angel (V_6): yes	4.0	3.0	4.0	(6,988)
Entrepreneurial climate index	131.4	108.2	126.6	(4,583)

[1] see table 1 for operationalization of entrepreneurial attitude variables of the entrepreneurial index
[2] the higher the value, the more favourable the entrepreneurial attitude

Data source: GEM population survey Germany June/July 2000 (N=7.034)

the regional areas differ from each other most. This is particularly true of the differences between federal states in eastern and western Germany, for opportunities are considered far less favourable in eastern Germany. Similarly, the fear of failure with one's own start-up and the prejudices against very successful entrepreneurs are much more widespread in eastern Germany than in western Germany. As table 2 shows, the proportion of those surveyed who demonstrate pro-entrepreneurial attitudes and activities is higher in western Germany than in eastern Germany for all six variables. This is without doubt well-founded statistical evidence for the thesis that climatic differences not only still exist between eastern and western Germany, but also that they tend to influence the difference in entrepreneurial ratios. Correspondingly, the value of the entrepreneurial climate index in western Germany is considerably higher than in eastern German regions.

Differences in attitude at the level of regional planning regions are particularly interesting from the point of view of economic geography. Specific climates may develop within such regions, and this is the scale at which most entrepreneurs think when they establish and use private and professional networks to realise their idea for a start-up. To back this up empirically, table 3 presents some results from the population survey on entrepreneurial attitudes and activities in seven selected German regional planning regions. These regions are among the most entrepreneurial regions in Germany, both in absolute and relative terms (cf. figure 3 and Fritsch/Niese also published in this book), and were represented with a relatively high number of cases in the GEM telephone interviews in the summer of 2000, meaning that the projection of the entrepreneurial climate from the interviewees to the region as a whole is not unrealistic.

Obviously, for most variables, these conurbations have proportions of pro-entrepreneurial features which are above the German national average. This is the case for all variables, without exception, for Munich and Stuttgart. Accordingly, regional planning regions which are dominated by large cities have a considerably more favourable entrepreneurial climate than other regions. The exception to this rule is the Oberes Elbtal/Osterzgebirge regional planning region, centred around Dresden, where the entrepreneurial climate is be-

Tab. 3: Entrepreneurial attitudes in selected German planning regions ("Raumordnungsregionen")

Statements[1]	Percentage of adult population with favourable entrepreneurial attitudes in selected planning regions ("Raumordnungsregionen", boundaries as of 1996)[2]								
	Berlin (N=324)	Hamburg (N=157)	Cologne (N=188)	Munich (N=214)	Upper Elb Valley/ East Ore Mountain (N=88)	Rhine-Main (N=223)	Stuttgart (N=226)	Other regions (N=5,614)	Staistical signifi-cance[3]
Similar standard of living (V_1): no	47.2	60.4	45.1	49.7	39.5	51.1	48.7	45.0	*
Fear of failure (V_2): no	59.6	65.9	60.5	60.0	52.0	60.1	56.2	55.3	
Good start-up opportunities (V_3): yes	29.6	35.5	37.1	50.3	30.3	44.0	41.5	28.0	***
Resentment against successful entrepreneurs (V_4): no	88.1	83.3	89.6	86.0	78.4	90.9	87.6	85.0	
Respect for entrepreneurs (V_5): yes	79.0	87.8	78.5	84.0	82.9	81.8	84.8	79.8	
Business Angel (V_6): yes	5.0	3.8	4.3	4.2	6.8	2.2	5.4	3.9	
Entrepreneurial climate index[1]	127.3	148.0	135.6	151.5	126.3	137.9	143.6	123.6	***

[1] see tab. 1 and section 2.2 for exact wording of the statements and for the definition of the entrepreneurial climate index

[2] shares higher than the national average are underlined, case numbers in brackets

[3] statistical differences between the seven planning regions for the dependent variables of column 1 (Chi-Square test); *: statistically significant at the 95 percent level of confidence, **: statistically significant at the 99 percent level of confidence, ***: statistically significant at the 99,9 percent level of confidence

Data source: GEM population survey Germany June/July 2000 (N=7.034)

low the national average and the individual variables, in particular those for the prejudices against very successful entrepreneurs and the expectations of a similar standard of living for the population as a whole (which is rather unlikely in an entrepreneurial society), are marked by poor values. The values for Berlin differ only slightly from those for Dresden. Subject to later analyses based on this data set, older references in literature to significant differences between eastern and western Germany regarding entrepreneurial attitudes among the population appear still to apply in the year 2000. But this must not be interpreted as the basis for a general and sustained lower level of entrepreneurial activity in eastern Germany, for attitudes can change. There are already indicators to support this; growth determinants for start-ups (including those of a regional nature) no longer differ fundamentally between eastern and western Germany, for example, but merely in scope, and start-ups in eastern Germany have considerably higher survival and growth rates than their western German counterparts (cf. Brixy 1999).

Analysis in chapter 2 showed that absolute and relative frequencies of start-ups differ regionally. There is still little of evidence of why that is the case and the role played in this explanation by the respective regional environment in comparison, for example, with the personal and firm-related aspects. The results of the Global Entrepreneurship Monitor for the first two years, 1999 and 2000, provide initial evidence, however, to suggest that the regional environment of the potential entrepreneur plays a more significant role than previously thought, for this is where the regionally specific attitudes and mentalities which influence the entrepreneurial tendencies in individual regions manifest themselves. As GEM continues, it will be possible to carry out intertemporal and interregional comparisons based on the same questionnaires, which will increase considerably the basis of information for providing explanations.

3. Research topics in economic geography

Neither entrepreneurial activities nor the later development of very young firms counts among the classical areas of research in economic geography (cf. Sternberg 2000b). The situation is similar within the discipline of business management studies, where the focus is also traditionally primarily on established companies. The tasks and areas of research in economic geography's research of entrepreneurship are mainly in the fields of developing and applying theories, empirical spatial research and spatial economic policy. An interdisciplinary approach is generally advisable – research into entrepreneurship based only on one single discipline does not produce adequate results – but this approach must take account of the comparative strengths of each individual discipline (cf. Schmude 1995). Within this framework of theory, empirical method and policy, spatially relevant subjects of entrepreneurial research can also be described, similar to the earlier outlined breakdown into determinants of entrepreneurial activity, entrepreneurial success and entrepreneurial promotion policies, an example of which is shown in table 3.

As far as theories are concerned, it should be emphasised that there is no theory derived from an economic geography point of view for start-ups. Growth and development theories developed and / or used by economic geography to date have not explicitly included the start-up sector, nor have economic theories in economic geography explicitly or adequately taken spatial characteristics into consideration (as part of external environmental factors, for example). This deficit in the existing theories may have been obstructive so far, but it does offer an opportunity and a challenge to at least partially close a significant gap in research. This may be done deductively, in view of the comparative strengths of (German) economic geography, but more likely inductively.

Tab. 4: Spatially relevant aspects of entrepreneurship research (selection)

	Variables with spatially divergent characteristics (selection)	Theses (selection)
Determinants of start-up activity		
– person-related, supply-oriented determinants	Motives, motivation, attitudes, competence	The motivation and attitude towards self-employment differs between immigrant and emigrant countries and explains the divergence in the frequency of entrepreneurial activities.
– environment and demand-oriented determinants	Locational factors, relevance of specific industries, firm size structure	Prosperous regions generate more start-ups, both in absolute and relative terms
Determinants of start-up success		
– factors related to the founder	Human capital resources, features of ego-centric networks	The availability of human capital and therefore the ability to develop egocentric networks varies between different types of region.
– factors related to the new firm	Industries, size	Start-ups tend to concentrate within certain regions in a small number of sectors of the economy (sectoral spatial start-up clusters)
– factors related to the environment	Features of the locality	The region as a location has an influence on the entrepreneurial dynamism, regardless of personal and commercial features
Start-up policies		
– design of new programmes	New programmes	Programmes have to be conceived taking regionally specific aspects into account if they are to be successful
– evaluation of existing programmes	Existing programmes	Similarly structured programmes generate differing results in different regions

There are very large areas of activity for empirical research of start-ups based on economic geography, some of which are stated in table 3 and in the concluding list. For most subject areas, research activities of this kind still require collation of primary data, as official statistics on start-ups do not exist and the majority of unofficial statistics do not allow any systematic form of regionalisation. This creates a challenge for economic geography; this is an opportunity to exploit a comparative strength over the economic sciences, for example. As Schmude/Leiner (1999) note, attention should be paid to avoid ecological and other kinds of erroneous conclusions which may occur when mixing data gathered at differing spatial levels.

Two examples are taken to illustrate the need for empirical work based on economic geography. Cyberspace is currently home to a particularly dynamic section of the start-up sector (in particular Internet start-ups) which is not adequately covered by official sources of figures. It is of scientific interest to examine the question of whether these start-ups, which at least technically are independent of traditional locational factors and spatial proximity, are actually more dispersed spatially in the sense of the “death of distance” (Cairncross 1997), or whether factors which, at first glance, appear to be non-economic, in the sense of spatial proximity advantages (which are often in combination with tacit knowledge and therefore *are* economic), nevertheless lead to a greater spatial concentration of such start-ups, as can be observed in the multimedia sector (Braczyk/Fuchs/Wolf 1999). From a spatial point of view, but also interesting in terms of promotional policy, is the thesis developed by Rosa/Scott (1999) for the United Kingdom according to which young start-ups are the most important source of strongly growing start-ups, rather than large companies or universities. This argumentation is not without a certain degree of plausibility, as persons with experience of launching start-ups should have better chances of successfully launching additional start-ups than somebody doing so for the first time. What is unknown (and this question has not been answered by secondary data either) is how and why entrepreneurs are spread spatially (more strongly or weakly concentrated than other incubator institutions?). The regional proportion of small firms in total, proven to be a variable with a strong correlation with high numbers of start-ups (cf. Bade/Nerlinger 2000 for western German regions), is not sufficient as an indicator because small firms are extremely heterogeneous in terms of their spin-off capabilities.

From the point of view of spatial economic policy, there is a need for economic geography to make its contribution to entrepreneurial research for two reasons in particular. First, there is a very large and still growing field of activity in the scientific evaluation of existing and the development of future programmes to promote start-ups. This is particularly true of empirically based research, as start-up promotion policies are often navigating through unknown territory, partly as a result of the lack of official statistics on start-ups, because they are based on insufficient knowledge of the numbers and needs of their target group, as well as their expectations of such programmes. In addition, as previously mentioned, there is no generally accepted theory for entrepreneurial activities, upon which the concept of policies to promote start-ups could be based.

Second, there is a clearly spatial component to this situation. Most programmes are initiated by individual federal states (e.g. the GO! initiative in North-Rhine/Westphalia) or municipalities (e.g. the majority of start-up centres) and therefore have a clear spatial objective and spatially specific target groups. Now, however, there is a national programme, "Exist", run by the Federal Ministry of Education and Research, which would like to stimulate start-ups in certain regions (in this case from universities, cf. Paffhausen da Cruz 1999). This demonstrates that the political world is increasingly, correctly, acknowledging a link between entrepreneurial activity and spatial factors. For the most part, however, there is still no answer to the question of which spatially differentiating factors influence entrepreneurial activities, how and where. Most studies of entrepreneurial activities to date have taken as their basis the national level and ignored regional factors. An additional reason why taking these factors into consideration is advisable is that unrealistic expectations are currently made of policies focussed on these start-ups. Even if current activities were to increase the number of start-ups, which is so far not generally the case, this does not mean that the expected effects on employment need necessarily follow. People only persuaded to launch a start-up by political encouragement may, under certain circumstances, be less successful than other self-employed people, as postulated, for example, by Fölster (2000) on the basis of empirical research in 24 Swedish Län.

If regional environmental factors do have a demonstrable influence on the generation and growth of start-ups, they should be explicitly incorporated into political initiatives to promote start-ups. A sensible approach would be "competitive federalism based on competition-oriented regional educational, infrastructural and economic policy", which may lead to specialisation on comparative advantages, which would be beneficial not only for the nation overall, but also for the majority of regions

(Reformkommission Soziale Marktwirtschaft 1999, p. 28). This need not necessarily mean that the regions currently winning the start-up race will remain the winners in the long term, and vice versa. The important aspect is that political strategies should be spatially concentrated and bundled – the question of how big such a region can, and must, be is then one for regional sciences to deal with.

In summary, the following scientifically and politically relevant research questions can be asked of economic geography:

- Which characteristics of the spatial economic environment influence the level of entrepreneurial activities and the success of start-ups?
- How can the considerable divergence in regional levels of entrepreneurial activities be explained?
- What form might an economic geographic argumentation take within the urgently needed new theories of entrepreneurial activities?
- What regional economic effects may realistically be expected from start-ups?
- What measures of economic and technological policy might increase regional start-up and survival rates?
- What are the conclusions of ex-post evaluations of existing programmes for promoting start-ups, with particular consideration of regional economic aspects?

Also of interest from an economic geography point of view is the link between the regional and the national level, remembering that the erroneous conclusions mentioned previously should be avoided. Still empirically untested in this connection is the hypothesis that the interregional distribution of entrepreneurial activities and the interregionally varying nature of framework conditions relevant to start-ups exerts an influence on the national level of entrepreneurial activities. The author will examine this issue in an interdisciplinary research project together with the Institute of Economics at the University of Lüneburg. The project will encompass the establishment of a Regional Entrepreneurship Monitor (REM) for German regions and translate the objectives of the above-mentioned Global Entrepreneurship Monitor (GEM) to a more in-depth, subnational level of observation. The combination of GEM and REM can be expected to generate an answer to this question of macro- and regionally economic relevance.

Economic geography should exploit the opportunities offered by the currently favourable political and economic framework conditions even more than before and leverage its comparative strengths over other disciplines – and be aware of comparative weaknesses. In the view of the author, the main strengths over micro- and macroeconomics, for example, are in investigations of the regional causes and effects of entrepreneurial levels and activities, based on primary data. The DFG Priority "Interdisciplinary Entrepreneurship Research", co-initiated by economic geography is aimed in precisely this direction, is active at exactly the right time and offers a worthwhile area of activity for this discipline.

References

Anselin, L., Varga, A. and Acs, Z. (1997): Local Geographic Spillovers between University Research and High Technology Innovations. Journal of Urban Economics 42, pp. 422–448.

Audretsch, D.B. (2000): Globalization, Entrepreneurship, and the Strategic Management of Regions. Paper presented at the Conference on Entrepreneurship, Firm Growth and Regional Development in the New Economic Geography. Trollhättan, Sweden, 15–17 June, 2000.

Audretsch, D.B. and Fritsch, M. (1994): The Geography of Firm Births in Germany. Regional Studies 28, pp. 359–365.

Bade, F.-J. and Nerlinger, E. (2000): The Spatial Distribution of New Technology-based Firms: Empirical Results for West-Germany. Papers in Regional Science 79, pp. 155–176.

Beise, M., Egeln, J., Gehrke, B., Legler, H., Licht, G., Machate-Weiß, V., Schasse, U. and Steincke, M. (1998): Zur regionalen Konzentration von Innovationspotentialen in Deutschland. Hannover, Mannheim (Studie von NIW u. ZEW im Auftrag des BMBF).

Bolton, R. and Westlund, H. (2000): Local Social Capital and Entrepreneurship. Paper Presented at the Conference on Entrepreneurship, Firm Growth and Regional Development in the New Economic Geography. Trollhättan, Sweden, 15–17 June, 2000.

Braczyk, H.J., Fuchs, G. and Wolf, H.G. (eds.) (1999): Multimedia and Regional Economic Restructuring. London.

Brixy, U. (1999): Die Rolle von Betriebsgründungen für die Arbeitsplatzdynamik. Eine räumliche Analyse für Ostdeutschland 1991–1996. – Beiträge zur Arbeitsmarkt- und Berufsforschung des IAB 230. Nürnberg.

Cairncross, F. (1997): The Death of Distance: How the Communications Revolution Will Change Our Lives. London.

Collinson, S. (2000): Knowledge Networks for Innovation in Small Scottish Software Firms. Entrepreneurship & Regional Development 12, pp. 217–244.

Department of Trade and Industry (1998): Our Competitive Future: Building the Knowledge Driven Economy. London. (see also www.dti.gov.uk/comp/competitive)

Fölster, S. (2000): Do Entrepreneurs Create Jobs? Small Business Economics 14, pp. 137–148.

Frick, S.; Lageman, B.; von Rosenbladt, B.; Voelzkow, H. and Welter, F. (1998): Möglichkeiten der Verbesserung des Umfeldes für Existenzgründer und Selbständige: Wege zu einer neuen Kultur der Selbständigkeit. – Untersuchungen des RWI 25. Essen.

Fritsch, M. (1997): New Firms and Regional Employment Change. Small Business Economics 9, pp. 437–448.

Fritsch, M. and Niese, M. (2000): Der Einfluß der Branchenstruktur auf das Gründungsgeschehen – Eine Analyse für die westdeutschen Raumordnungsregionen 1983–1997. Geographische Zeitschrift 88, pp. 234–250.

Funke, M. and Niebuhr, A. (2000): Spatial R&D-Spillovers and Economic Growth – Evidence from West Germany. – HWWA Discussion Paper 98. Hamburg.

Hendry, C., Brown, J. and Defillippi, R. (2000): Regional Clustering of High-technology-based Firms: Opto-electronics in three Countries. Regional Studies 34, pp. 129–144.

Institut für Mittelstandsforschung (2000): Wissenschaftliche Begleitforschung 1998/1999 und Würdigung der Gründungs-Offensive Nordrhein-Westfalen “GO” NRW. – IfM-Materialien 142. Bonn.

Johannisson, B. (2000): Modernising the Industrial District – Rejuvenation or Managerial Colonisation. M. Taylor and E. Vatne (eds): The Networked Firm in a Global World: Small Firms in New Environments. Aldershot, pp. 283–309.

Klandt, H. (1984): Aktivität und Erfolg des Unternehmensgründers. Bergisch-Gladbach.

Lageman, B. and Löbbe, U. (1999): Kleine und mittlere Unternehmen im sektoralen Strukturwandel. – Untersuchungen des RWI, 27. Essen.

Lessat, V., Hemer, J., Eckerle, T., Kulicke, M., Licht, G. and Merlinger, E. (1999): Beteiligungskapital und technologieorientierte Unternehmensgründung: Markt – Finanzierung – Rahmenbedingungen. Wiesbaden.

Lompe, K., Kehlbeck, H. Schirmacher, A. and Warnecke, D. (1998): Existenzgründungen, Risikokapital und Region. Baden-Baden.

May-Strobl, E. (1999): Gründungsförderung muß praktische Hilfe für den Gründer sein. In: Handelsblatt v. 5./6. März 1999.

Moßig, I. (2000): Lokale Spin-off-Gründungen als Ursache räumlicher Branchencluster. Das Beispiel der deutschen Verpackungsmaschinenbau-Industrie. Geographische Zeitschrift 88, pp. 220–233.

Nerlinger, E. (1998): Standorte und Entwicklung junger innovativer Unternehmen. Empirische Ergebnisse für West-Deutschland. – ZEW Wirtschaftsanalysen 27. Baden-Baden.

Paffhausen da Cruz, U. (1999): Die Rolle der Universitäten in Deutschland. Deutsch-Japanischer Kooperationsrat für Hochtechnologie und Umwelttechnik (ed.): Beiträge der Industrie, der Universitäten und Forschungseinrichtungen zur Unterstützung und Förderung von Existenzgründungen in Hochtechnologieund Umwelttechnik. Bonn, pp. 115–142.

Preisendörfer, P. (1996): Gründungsforschung im Überblick: Themen, Theorien und Befunde. P. Preisendörfer (ed.): Prozesse der Neugründung von Betrieben in Ostdeutschland. – Rostocker Beiträge zur Regional- und Strukturforschung 2. Rostock., pp. 7–29

Prevezer, M. (1998): Clustering in Biotechnology in the USA. G.M.P. Swann, M. Prevezer and D. Stout (eds): The Dynamics of Industrial Clustering. Oxford, New York, pp. 124–193.

Reformkommission Soziale Marktwirtschaft (1999): Die Renaissance der Selbständigkeit. Pioneer Square Deutschland. Gütersloh.

Reynolds, P.; Hay, M. and Camp, M.C. (1999): Global Entrepreneurship Monitor. 1999 Executive Report. Kansas City (see also www.entreworld.org).

Reynolds, P.; Storey, D.J. and Westhead, P. (1994): Cross-national Comparisons of the Variation in New Firm Formation Rates. Regional Studies 28, pp. 443–456.

Roland Berger & Partner GmbH (1998): Untersuchung der Wirkungsmechanismen von Kompetenzzentren am Beispiel ausgewählter Technologieregionen in Deutschland. München (Gutachten für den BMBF).

Rosa, P. and Scott, M. (1999): Entrepreneurial Diversification, Business-cluster Formation, and Growth. Environment and Planning C: Government and Policy 17, pp. 527–547.

Saxenian, A. (1996): Regional Advantage. Culture and Competition in Silicon Valley and Route 128. Cambridge, London (2nd Ed.).

Schamp, E.W. (2000): Vernetzte Produktion. Industriegeographie aus institutioneller Perspektive. Darmstadt.

Schmude, J. (Hg., 1995): Neue Unternehmen. Interdisziplinäre Beiträge zur Gründungsforschung. – Wirtschaftswissenschaftliche Beiträge 108. Heidelberg.

Schmude, J. and Leiner, R. (1999): Zur Messung des Unternehmensgründungsgeschehens: theoretische Überlegungen und empirische Befunde. L. v. Rosenstiel und T. Lang-von Wins (eds): Existenzgründung und Unternehmertum. Stuttgart, pp. 109–128.

Schutjens, V.A.J.M. and Wever, E. (2000): Determinants of New Firm Success. Papers in Regional Science 79, pp. 135–154.

Sternberg, R. (1998): Technologiepolitik und High-Tech Regionen – ein internationaler Vergleich. – Wirtschaftsgeographie 7. Münster, Hamburg (2nd revised edition).

Sternberg, R. (2000a): Entrepreneurship in Deutschland. Das Gründungsgeschehen im internationalen Vergleich. Länderbericht Deutschland 1999 zum Global Entrepreneurship Monitor. Berlin.

Sternberg, R. (2000b): State-of-the Art and Perspectives of Economic Geography – Taking Stock from a German Point of View. GeoJournal:50, pp. 25–36.

Sternberg, R., Behrendt, H., Seeger, H. and Tamásy, C. (1997): Bilanz eines Booms – Wirkungsanalyse von Technologie- und Gründerzentren in Deutschland. Dortmund (2nd revised edition).

Sternberg, R. and Tamásy, C. (1999): Munich as Germany's No.1 High Technology Region – Empirical Evidence, Theoretical Explanations and the Role of Small Firm/Large Firm Relationships. Regional Studies 33, pp. 367–377

Torre, A. and Gilly, J.-P. (2000): On the Analytical Dimension of Proximity Dynamics. Regional Studies 34, pp. 169–180.

Wennekers, S. and Thurik, R. (1999): Linking Entrepreneurship and Economic Growth. In: Small Business Economics 13, pp. 27–55.

Zucker, L.G.; Darby, M.R. and Brewer, M.B. (1994): Intellectual Capital and the Birth of U.S. Biotechnology Enterprises. – NBER Working Paper Series 4653. Cambridge/Ma..

Corresponding address:
Prof. Dr. Rolf Sternberg
Institute of Economic and Social Geography, University of Cologne
Albertus-Magnus-Platz
50923 Cologne
Phone: ++49-221-470-2372
Fax: ++49-221-470-5009
sternberg@wiso.uni-koeln.de
www.wiso.uni-koeln.de/wigeo/index.html

(First published in Geographische Zeitschrift vol. 88/ 2000 (3 + 4), pp. 199–219

Geographische Zeitschrift, Special issue 2004

Culture, geography and the diacritical practice of 'Oriental entrepreneurs'

MARC BOECKLER, Eichstätt

Summary

Edward Said's 'orientalism', the globalization discourse and the 'cultural turn' within the social sciences constitute important challenges to geographers who focus their research on the 'Orient'. Having traditionally positioned themselves within an empirically orientated subdiscipline, scholars need to develop approaches which go beyond essentialism, methodological territorialism and scientific realism. This paper advances a perspective of dealing with cultural processes which starts from cultural hybridity as a radicalized way of viewing the social world and links this with an empirically manageable notion of culture through the introduction of transculturality as a heuristic intermediate concept. In this view culture is understood metaphorically as a daily practice of separating, differentiating and distancing, that is, as a *diacritical practice*. Syrian entrepreneurs are in this vein conceptualized as hybrid actors living in the betweenness of different worlds, the empirical emphasis being put on the entrepreneurial practice of symbolic and material 'bordering' as well as imaginary and bodily territorialization. In the 'global age' it is to a large extent through their diacritical practice that Syrian entrepreneurs contribute to the construction of the 'Orient', a process which is considered to be a 'consequence of modernity'.

1. Disorientation/Reorientation

The entrepreneur has always been a fascinating spectre spreading both fear and hope. Interwoven with Occidental narratives of modernity the theoretical figure of the entrepreneur insists rather incisively on the paradoxical nature of modern culture. By simultaneously creating 'new' and destroying 'old' newness, producing rules as well as breaking them, entrepreneurs have become the celebrated heroes of economic development and the villains of social inequality. They are much-courted favourites of neoliberal economic policies and, because of their propensity to profit from market imperfections, secret enemies of neoclassical economic theories. But this paper is not about entrepreneurs in economic theory. With the help of 'oriental entrepreneurs' it aims to contribute to global geography after Orientalism.

Modern social science invented the entrepreneur as a discursive category to assure itself of its own rational foundation in opposition to the absent 'Oriental other'. Max Weber (1988, pp. 1–8) made it perfectly clear: "Nur auf dem Boden des Okzident" – only on Occidental territory – could capitalism develop as a distinct spirit embodied in outstanding individuals gifted with an exceptional talent for rational action and long-term planning, and prepared to cut primordial ties in order to rigidly separate family and business. By stressing his innovative capacity, Schumpeter (1934) additionally provided the entrepreneur with the individualistic quality of a single hero who is swimming against the tide of the masses. On 'Oriental territory' however, imaginative geographies of economies left no room for a powerful Faustian 'I' within the undifferentiated collective 'We' of faceless Oriental masses. Economic subjects were represented as idle, lazy, passive, inflexible. Equipped with an inclination to imitate and exploit, they were imagined as looking for high and quick profits embedded in family loyalties, loyalties which dominated economic rationality (see Banse 1926, pp. 7–19; Sayigh 1957). Even more sophisticated political economic accounts of the Middle East put it briefly and succinctly: "The Oriental entrepreneur is the Anti-Schumpeter" (Beblawi 1987, p. 52).

With Jacques Derrida's critique of 'logocentrism' and Edward Said's deconstruction of Orientalism the epistemological self-delusion of representational *disorientation* became obvious: binary thinking operates to exclude the Oriental entrepreneur as the Western entrepreneur's other. But the two sides remain connected through temporalized and spatialized différance (Derrida 1972): the 'Oriental entrepreneur' is the deferred (différé) and different (différent) 'Western entrepreneur'. More important: in order to maintain the illusionary objectivism of the real Other, modern geography conceptually put an end to the ceaseless play of meaning by drawing a veil over the intricate interdependence between Orient and Occident. The veil consisted in dramatic boundaries created to enclose symbolic territories of the cultural different. Only on the basis of this territorial opposition could the powerful colonial study produce knowledge about 'Orientals' and "orientalize the Orient" (Said 1995, p. 49).

As important as deconstructivist work has been for propelling an overdue reflexive and constructivist cultural turn throughout social science, pure deconstructivism has of late become a somewhat tiring version of disorientation itself. Following Latour's (1998, p. 280) diagnosis that we have shifted from cold, detached and certainty desiring science to warm, involving and uncertainty producing research, we must state that the self-inspection of modernist discourses has successfully exorcized any *science of* the Orient. However, it has not been nearly as successful in triggering *research with* the Orient. In this respect some sort of *reorientation* might be necessary. It is time to literally go back to the 'real' places where people carry out activities and construct meaningful temporary social realities in the constant flow of materially interwoven practices. German 'Kulturgeographie'[1], in particular, looks back at a strong and fruitful tradition of detailed and attentive "thick descriptions" (Geertz 1973) of various aspects of daily life in the Middle East and North Africa (see Kopp 1999). But it has not yet made creative use of the critique of Orientalism and until recently held on to the modernist meta-narrative of a natural world-order composed of distinct culturally defined territories.

In this paper I would like to reorient research with the Orient in a threefold manner. First, with the concept of 'Entterritorialisierung' I propose a postterritorial way of thinking about spatial configurations of social practices in a global age; second, culture will be introduced as diacritical practice, a conceptualization of culture in a neither totalizing nor dissolving fashion that focusses on boundary processes. Finally, I will try to befriend and engage with 'Syrian entrepreneurs', who are considered to be at once transterritorial hybrids situated between-the-worlds as well as powerful actors inscribing material and symbolic boundaries in the Syrian economic landscape.[2] This engagement has two aims. First, I seek to exemplify my theoretical argument by illustrating the 'Oriental production of the Orient'. Second, I intend to cautiously approach the modern entrepreneur's excluded Other with the unconcealed geo-romantic aim of contributing a particular narrative to what Doreen Massey has termed "real differences" of a truly spatialized globalization (see Massey 1999a, p. 13). By telling the stories of Oriental entre-

1 In a few selected cases, I decided to stick with the German original. English translations of German terms and texts are mine.

2 This paper draws on interview and ethnographic material gathered from 1996 to 1999. The author acknowledges financial support from the interdisciplinary Bavarian research network FORAREA.

preneurs, I hope to recount local stories of a global narrative in which spatial differences are no longer temporalized into simple inside-outside oppositions via theories of evolution, development, modernization and so on, but where space is conceptualized as "the sphere of the possibility of the existence of multiplicity" (Massey 1999b, p. 2).

At last a short warning: the paper does not mimic the rigor of so-called hard science, but deliberately tries to develop a more playful style with a 'positive eclecticism' as a core element, i.e. optimistically making use of the strengths of different approaches without denouncing their weaknesses and shortcomings.

2. *Entterritorialisierung*

For quite some time now 'surface thinking' has lost its plausibility and the world's spherical form imposes itself: the 'globe' heralds a new age in which our very life circumstances have changed fundamentally. Giddens (1990, p. 64) theorized these changes as "intensifying worldwide social relations", "time-space-distanciation", "interaction across distance", "connections of presence and absence" and Martin Albrow took the geometrical metaphor literally, asking rhetorically: "Can it be anything other than the 'Global Age'?" (Albrow 1996, p. 80). Within view of the shoulders of these giants, is it blasphemy to simply ask again? What is so revolutionary about globalization? Quite a few centuries before Western discourse identified transnational corporations as engines of economic globalization, traders from Aleppo organized their business through a global web of partners and branch offices. And speaking of political denationalization, it has to be pointed out that after a few centuries under Ottoman domination and French colonialism the invention of Syria as a nation state was a genuine transnational project carried out by international forces within French and British borders. With respect to the transcultural qualities of globalization, it is easy to remember that particularly in Syria "for at least five thousand years (...) different cultural currents have met", the region ever since having been an interface between "Orientalization and Occidentalization" (Wirth 1971, p. 137, 139). So what? Where is the revolutionary novelty preached by the high priests of globalization? Read as a simple substantive and descriptive diagnosis of recent social processes, the 'everything-connects-to-everything' topos does little to go beyond the "Ausweitung der Interdependenzketten" – the expanding chains of interdependencies identified by Norbert Elias (1976) even for pre-modern ages. In contrast to this, the global age develops its revolutionary power first and foremost through a fundamental conceptual shift. It is the reflexive extra effort of 'relational thinking' that lifts the conceptual veil of modern territorialization, an effort which reveals the world as a globe.

Read relationally, globalization has been unravelling the modern sociospatial order of a culturally territorialized world. I would like to term this double challenge of substantive processes and conceptual perspectives 'Entterritorialisierung'. I prefer Entterritorialisierung (Entterritorialization)[3] to globalization and deterritorialization, given that 'globalization' has lately become 'economicized' in a rather one-dimensional manner and 'deterritorialization' is usually either reduced to the declining importance of national territoriality or it falls victim to a rhizomatic physicism as de- and

3 The German expression 'Entterritorialisierung' may be read as 'Entterritorialization'. The prefix 'ent-' has three different meanings in the German language: It derives from 'entgegen' which means 'against/the opposite', and is used in this initial sense as 'to undo' something; however, more often it is applied in the wider sense of 'removing', e.g. 'entnehmen' (take out) as compared to 'nehmen' (take); in its third connotation 'ent-' emphasizes and reinforces an act of removal, e.g. 'fliehen' (flee, escape) and 'entfliehen' (flee, escape). Following this, I understand Entterritorialization as the reinforced dismissal and undoing of territorialization.

reterritorialization in topological processes (see Deleuze/Guattari 1986). But Entterritorialisierung is not nomadology; it is rather a provisional collection of a few methodological preferences:

- First, Entterritorialisierung signifies a consciousness that rests on the "extensive awareness of the world as a whole" (Robertson 1992, p. 78). However, this globality is neither an argument for the true existence of a single world society nor a denial of existing social disparities. It is a conceptual argument, that re-entangles the previously atomized individual into the "Gesamtzusammenhang des bewegten Menschengeflechts" (Elias 1987, p. 86) – the total nexus of a constantly changing web of relational human beings.
- Second, after having been veiled by a powerful "methodological nationalism" for a long time (Smith 1979, p. 191), the vantage point of the 'world-as-a-whole figuration' makes the fundamental transterritorial connectivity visible again which underlies all processes of territorial order-making. It seems to be perfectly clear: modern science followed an implicit nationalism and developed a conceptual territorialism where all aspects of social reality, ranging from the power container of nation states to the bodily self-contained autonomous subject were represented as bounded and homogeneous. It is relational thinking that recognizes the contingencies of boundary processes. Speaking in terms of systems theory, we become aware of the 're-entry' of the distinction of inside-outside into the process of inclusion and exclusion itself (see Luhman 1997). As a consequence, a general relational geography has to consider both, the blurring of boundaries by focussing on multiple internal relations as well as the continuing production of external differentiations. What we are now able to see is the intricate simultaneity of connecting and separating (see Zierhofer 1999, p. 9f.). Trying to escape the "territorial trap" (Agnew 1994), Entterritorialisierung reworks conceptual territorialism to a post-territorial focus on boundary processes, whose outcome might indeed resemble an isomorphism of culture and national territory. However, it is just as likely that sociality will be produced in disjunctive global ethnoscapes (Appadurai 1990), in local sociospheres (Albrow 1997), along travelling routes (Clifford 1997) or in diasporic social relations (Hall 1990).
- Third, if we accept – without having logical proof for it – the plausible assumption that the perspectival shift corresponds to a transition of social reality from a modern to a global age, we may understand Entterritorialisierung as one of those multitudinous unintended consequences in the sense of Ulrich Beck's 'reflexive modernization' (Beck 1986). 'Global modernity' emerges in the process of a worldwide diffusion of what Giddens (1990, pp. 55–78) conceived as 'institutional dimensions of modernity' (industrialism, capitalism, surveillance, military power), facilitated by technological innovations in transportation, information and communication systems. But Entterritorialisierung is not a simple consequence of worldwide technological or institutional diffusion. Rather, it comprises the fundamental risk and uncertainty produced by the newly developed single global techno-discursive frame of reference which transcends conceptual territorialism and discloses the mutual inscription and interdependencies previously veiled by dramatic boundaries. The epistemic risk rests precisely in the unpleasant finding that the neatly separated metropolitan and peripheral, colonialized and colonizing cultures and powers have from the very beginning been drawn together in an inseparable web of mutual interrelatedness. In this respect Entterritorialisierung forms an alliance with "the 'post-colonial' as an episteme-in-formation" (Hall 1996, p. 255). It attacks the very self-constructed foundations of West-

ern hegemony over modernity, which dissolves into stories of heterogeneous global modernities, stories simultaneously told by multitudinous local narratives that blend into a colourful, undifferentiated, unequal and contradictory noise of a plentitude of different spatialities.

- Fourth and most importantly, beyond conceptual territorialism and with substantive globalization processes in view, Entterritorialisierung develops an empirical sensitivity for the fundamental renegotiation of the spatiality of social relations in a global age. Entterritorialisierung then considers places as 'world-places', literally as 'cosmopolis', always cosmos, and always polis, produced by interrelations and constituted in and through social practices. To be precise, with respect to space, place is not to be found in space, but is a nexus of mediation, "a means by which space is produced as a *plentitude of different relations*" (Thrift 1999, p. 310). With Schatzki (1996, ch. 4), one should insist on the annex yet again: space produced in and through social practices. The relations which practically constitute place are not reduced to precontained local territories, but are distancially undetermined. Benno Werlen (1997) has coined the term "Welt-Bindung", which comprises the opening up of places to the world as well as the local appropriation of the world in practices as different aspects of world-places: "Space, then, is the product of the intricacies and complexities, the intertwinings and the non-interlockings, of relations, from the unimaginably cosmic to the intimately tiny" (Massey 1999b, p. 8).

Selectively following Norbert Elias, that is not in his substantive theory of an evolutionary civilization process, but in his non-modern theory of relational society, world-places become 'non-territorial local practice-figurations'. Elias was to free the 'Ich im verschlossenen Gehäuse' (Elias 1996, p. 141) – the "I in the locked casing" – from the conceptual prison and to open up the "homo clausus" to the "homines aperti" (ibid., p. 135). Social reality is not made up either of autonomous subjects nor of bounded societies, but of interconnections, interdependencies, interrelations of constantly moving human beings. In short, Elias conceptionalizes human beings as relational processes (ibid., p. 127), with figurations constituting these fluid patterns of interrelated social practices (ibid. 142). For Elias, fluctuating power asymmetries, "fluktuierende Machtbalancen" (ibid., p. 143), are relationally intrinsic to figurational flows. Following from this, in the global age places may be conceived of as always plurilocally ordered, as 'entterritorialized' flows of practice-figurations.

3. Hybridity, transculturation and culture as diacritical practice

> "I have been arguing that the 'Orient' is itself a constituted entity, and the notion that there are geographical spaces with indigenous, radically 'different' inhabitants who can be defined on the basis of some religion, culture or racial essence proper to that geographical space is equally a highly debatable idea" (Said 1995, p. 322).

Where is the place of culture in Entterritorialisierung? Culture was previously understood as a fixed 'thing', composed of particular meanings, devices and practices, and was, even as a 'way of life', thought to pre-date the individual and instruct individual action. The "cultural stuff" (Barth 1969, pp. 10–15) was conceived as stable, territorially bounded and internally homogeneous. Accordingly, the world as a whole resembled a mosaic of different cultures with an Occidental centre and a multitude of different Others. Binding cultural purity to 'blood ties' and linking it with territorial homogeneity can have disastrous consequences, as the German past has demonstrated. This notwithstanding, a territorialized notion of culture was the foundational principle of German 'Kulturgeographie' in the late 19th and early 20th century (Schlüter 1906, p. 28). This may explain why the discipline finds it so difficult to part with this notion of culture once and for all. But again, it is the double attack of epistemological and substantive developments, of cultural turn and globalization that ripped this concept of culture to pieces and demands different notions of culture.

3.1 Hybridity

During the last decades new concepts of culture emerged that stressed the mobility, fluidity, rootlessness but routedness of culture as practice. To date the most radical and creative critique of essentializing and homogenizing notions of culture has been offered by the concept of 'hybridity'. Quite frequently, hybridity is misunderstood as 'mixture' within a wider stream of multiculturalism and is used synonymously with other metaphors like syncretism, creolization, and métissage. But cultural mixture is not a particularly innovative account, neither as theoretical concept nor as empirical statement: Franz Boas, the German-American founder of anti-racist cultural anthropology, already stressed the mixed character of traditional societies at the beginning of the last century. In his words

> "the cultural form may become a kaleidoscopic picture of miscellaneous traits that, however, are remodelled according to the changing spiritual background that pervades the culture and that transforms the mosaic into an organic whole" (Boas 1927, p. 7, cited in Friedman 1994, p. 75).

However, hybridity does not express mixture but non-identity. Where Boas asked for processes of integration, hybridity radically criticizes the thinking in terms of organic wholes, of fixed and stable, homogeneous, territorially bounded, purified concepts of cultural totalities in general. Hybridity displaces the thinking in taken-for-granted inside-outside dichotomies, replacing it with the diverse multiplicity of relationally produced internal differences. In so doing, hybridity marks the 'derridaesk' move from difference to différance (Hall 1992). Like meaning, which is caught in the incomplete chain of signification, culture and identity are always deferred and differed, never fixed, stable, pure, but tossed around by the ceaseless play of differences. The place of culture and identity is consequently the interstitial space, the 'in-between space', where the 'in-between' is not read as between the one and the other, but as Third Space in the sense of neither the one nor the other (Bhabha 1994, p. 36f.), as the impossible space of the simultaneously 'not yet' and 'no longer', 'not here' and 'not there'.

Hybridity has been much criticized lately. First, in counterpoising hybridity to essentializing conceptions of culture, the hype of hybridity ironically reproduced binary structures. Second, hybridity is being seen as a new version of cultural imperialism, in which the historically situated US debate on multiculturalism is de-contextually universalized. In this way Pierre Bourdieu and Loïc Wacquant (1999, p. 43) criticize "'postmodernism' which, founded on an eclectic, syncretic and, most often, dehistoricized and highly approximate re-reading of a platoon of French and German authors, is in the process of being imposed in its American form upon Europeans themselves." Third, according to some critics, hybridity combines a distinguished intelectualocentrism with a gaping empirical void, because only the academic representation of practice is hybrid, never the daily life of an actor, no matter how marginalized she might be (Friedman 1999). There is some legitimacy to this criticism, but critics develop no alternative to the politics of difference supported by theories of hybridity. Entterritorialisierung accepts hybridity as a radicalized reading of the world and as an attentive escort to cultural research, as an indispensable companion warning of essentializing representations, otherings and representational exclusions. Besides, it was in the 'in-between space' of hybridity where the most innovative contributions to social and cultural science were developed in the late decades of the 20th century, brought forward by "halfies", the label Lila Abu-Lughod (1991, p. 137) chose to depict herself ironically as occidentally contaminated Occidental academic of non-Occidental origin. Later I will grant the same epistemologically privileged position to the Syrian entrepreneur.

3.2 Transculturality

First and foremost, hybridity points to the fact that relational geographies are a 'boundary project' as well and that any identification is accompanied by boundary processes. Therefore the task will be to keep "the process of boundary construction in view" (Pratt 1999, p. 156). But it should be added that it will be far more difficult to cope with this work with respect to Bavarian dairy farmers, Pittsburgh steel workers or Syrian entrepreneurs than with Philippine workers in Canada or Indian computer experts in the Emirates. And there is another problem. Theories of hybridity remain silent when it comes to powerful inscriptions of stable and fixed cultural identities. In Germany, for instance, political programs of cultural integration rest on the inscription of presumed homogenous Turkish origins and follow conventional ideas of purified German national culture. It is impossible to grasp adequately the everyday life of 'türkische Deutschländerinnen' (Zaimoglu 1998) without taking into account the prevalence of an old notion of culture in the political arena.

It is at this point that Wolfgang Welsch's concept of 'transculturality' is helpful. Here transculturality is not used in Fernando Ortiz' (1940) sense of transculturation, where tobacco and sugar symbolized two different and unequal, but intermingling cultures of Cuba, but as a position in-between two different conceptualizations of culture. Welsch criticizes the old island model of bounded culture in its different manifestations either as 'single culture' (one island), 'multi-culture' (single islands as part of an archipelago) or 'cross-culture' (island hopping). In contrast to this, transculturality conceives culture as hybrid in the above sketched sense, but additionally signals a double situation of transition, referring

> "to a transition, or to a phase in a process of transition. It's a temporary diagnosis. (...) The process of transition obviously implies two moments: the ongoing existence of single cultures (or of an old understanding of culture's form) and the shift to a new, transcultural form of cultures" (Welsch 1999, p. 208).

In other words, transculturality not only sees heterogeneity and multiplicity as a new concept of culture but takes into account the existence of multiple and heterogeneous concepts of culture as well. As a strategic position 'transculturality' helps to keep in view everyday practices of essentialization.

3.3 Culture as diacritical practice

The idea of culture was invented in 18th century Europe. At a time when God was being replaced by man, culture was the very concept that announced and executed the task of order-making by a threefold process of separation: culture as the sum of human accomplishments was separated from nature, newly built societies were reorganized into a group of those that possessed culture as high culture and those that did not, and finally the world as a whole was arranged in a territorial mosaic of different and finely separated cultures, where, of course, some cultures were better than others (Fisch 1992). With the transition to a global age this modern cultural world order has been challenged and become heavily disorganized in all its dimensions. Whereas advocates of a cultural turn welcome the colourful creative disorder, old border patrol agents have mounted the stage shouting for the militarization and reinforcement of disappearing borders. Culture wars take place at different scales, within societies, within the academic arena and are also propagated in a jihadist sense on a global level as "clash of civilizations" (Huntington 1996). Instead of evaluating good or bad cultures or developing better or worse concepts of culture, I would like to centre the metaphor of culture on the very core of the concept that is order-making by separating:

> "The cultural contribution to nature (...) does not consist in the provision of an artificially produced and reproduced total and normatively regulated way of life, but in the continuous impulse to differentiate, separate, divide, classify: to the conjuring up of new meanings through diacritical practice" (Bauman 1997, p. 133).

'Culture' then signifies neither a distinct way of life nor a structured totality, nor does it serve cultural geography as an 'object' of study or as a 'factor' explaining particular manifestations of behaviour. As an abundant metaphor culture simply draws the attention to everyday boundary processes in different fields and dimensions of human practice. On the basis of a relational approach and with the perspective of Entterritorialisierung, culture symbolizes the continuing insertion of contingent separations into a fundamentally interconnected world – or with Martin Heidegger (1990, p. 149) in a somewhat more philosophical tone: "die Grenze ist jenes, von woher etwas *sein Wesen beginnt*" – the boundary is that, from which something begins its presencing, and one could add, where places are spaced and spaces opened up (ibid., p. 151). In short, culture is 'diacritical practice'.

In linguistics the Greek diakritikós is used to denote marks that differentiate homographs. This is of particular importance in Arabic script where 28 graphemes rest on 18 different graphs, of which 13 are homographic and can only be differentiated through diacritical marks. Arabic linguistics is very precise here and speaks of *naqt* as the practice of setting dots above and under the graph and of *i'jam* as the overall aim of this practice to 'resolve the lack of clarity'. Nevertheless, from a relational perspective it is obvious: the separated connectivity of sameness continues to exist in the homographs. As denoted above, the reflexive extra effort of 'relational thinking' enables recognition of the intricate simultaneity of connecting and separating. In this respect geographies of Entterritorialisierung are hyphenated geographies, like the hyphen that joins two words or parts of a word and at the same time separates them, geographies of Entterritorialisierung focus on the diacritical practice of separating but keep the underlying connectivity in view.

It is not necessary to follow Bourdieu's (1979, p. 553) essentialism all the way, when he argues that "le sens de la distinction, discretio" develops a "horreur viscérale, dégout absolu", deadly horror and absolute disgust, at all ambivalences that contradicts the "ordre social fait corps", the incorporated social order, to accept the observation that distinction as everyday boundary process is inherent in social practices. In Elias' (1976) grand theory the process of civilization is strongly influenced by the constant effort of distinction and distanciation. This process of distinguishing boundary construction can be conceived of as an always present process of negotiation and inscription of meaning and identity - no matter how impossible these attempts to 'resolve the lack of clarity' may be. As mentioned above, it is important to recall that diacritical practices are characterized by inherent power asymmetries that unequally distribute the rights of boundary construction and border crossing.

Furthermore, diacritical practices draw on 'practice theories' as the currently most promising answer to postmodern challenges within social theory. As practice accounts follow an irreducible ontology, starting with the assumption "that interaction is all there is" (Law 1992, p. 380), they do not solve, but simply ignore, modernist dichotomies of actor and structure, micro and macro. Practices then are conceived of, minimalized, as arrays of activity. If 'posthuman' practice theories are disregarded, practices can be understood as an embodied set of human activities, mediated by artefacts, hybrids and natural objects and organized around practical understandings. Actors and actions, knowledge and structure are not only embedded within practices, but are in a Wittgensteinian sense constituted in and through practices. In this respect practice thinking is radically contextual. From the perspective of Entterritorialisierung context is conceptualized as total nexus of interconnected practices, therefore practices decide the context and context is in this very sense unlimited. Or to paraphrase Derrida: 'il n'y a pas de hors *con*-texte' (see Derrida 1988) – there is nothing outside context. Order is therefore not considered as patterned regularity, but as a contextual arrangement of people, artefacts and things.

In order to develop a decentered idea of the actor in an unlimited context, I would like to bravely ignore Giddens' (1987) latter harsh rejection of poststructuralism and go back to early structuration theory. There, "structuration, or as Derrida puts it, 'the structuring of structure'" (Giddens 1979, p. 61) still followed "a threefold connotation of différance" (ibid., p. 46). The memory traces that are instantiated as structure through action within structuration (see Giddens 1984, p. 17) must clearly be read as temporalized and spaced traces in Derrida's sense, as the traces structuration has left behind at the intersection of absence and presence. If the notion of an actor is additionally enriched with Elias' figurational flows, this actor can be conceived of as an open, conflictually multiplied and fragmented bundle of contextually changing relations. This provides for a reasonably acceptable notion of the theoretical figure of an actor in diacritical practice that is not to be reduced to an autonomous subject possessing an uncontested single and bounded identity or as individualistic antagonist of collective structures. With this image of an actor in mind, I would now like to cautiously approach diacritical practices of Syrian entrepreneurs.

4. Syrian entrepreneurs and diacritical practice

In the preceding section I sketched a playful approach to the current contradictions of cultural efforts to order and classify. Starting from (1) a radicalized world-view (*hybrid culture*) and (2) the strategic position of *transculturality*, I advanced an understanding of culture as (3) *diacritical practice*. This section applies this general conceptual framework to the figure of the industrial entrepreneur and, in a second step, exemplifies diacritical practices of Syrian entrepreneurs from a perspective of Entterritorialisierung. These examples raise the question of how the structuration of business firms contributes to the production of plurilocally ordered and deterritorialized places as local practice-figurations. Here, 'diacritical practices' regionalize the world, territorialize places and produce unlimited local contexts.

4.1 The entrepreneurial actor

(1) For the purpose of this paper, the figure of the entrepreneur is an ideal starting point for a hybrid reading of socially constructed realities. Not unlike the epistemologically privileged migrant intellectuals dominating the discourse of postcolonialism, the Syrian entrepreneur occupies an elevated position *in-between-the-worlds:*

- With the daily transformation of material nature into commodified culture, entrepreneurial practice demonstrates the relational simultaneity of the purifying separation and the translating connection of the spheres of nature and culture (Latour 1993).
- Through the structuration of industrial plants entrepreneurs open multiple translocal 'landscapes' ('Syrian-French' capital, workers from Aleppo, 'Japanese' management models, 'German' machinery) and, with the positioning of the business firm between these symbolic worlds, serve as a constant reminder that culture is always in the making, never completed, never finished.
- And, finally, Syrian entrepreneurs display everyday knowledge acquired through translocal biographies – spending long periods of education and work abroad, above

all in Europe, the US and Latin America. As 'Occidentally contaminated non-Occidentals' they radically question the idea of culture as something homogeneous and locally bounded.

(2) Historically, the discursive figure of the entrepreneur is deeply rooted in Occidental thinking. The category of the 'Syrian entrepreneur' becomes legitimate only within a global techno-discursive frame of reference. As a 'consequence of modernity', the 'Syrian entrepreneur' then evolves into a *transcultural category*, serving as a reminder of the deeply intertwined processes of modernization and colonialization and the dissolution of the world as a mosaic of territorialized cultures:

- In its symbolic dimension the figure of the entrepreneur represents bourgeois virtues: systematic organization of professional and private life, economic rationality, bookkeeping, thriftiness and meaningful use of time as pillars of moral fulfilment (Hansen 1992). 20th century social science reconstructed the entrepreneur as an icon of Western modernity, in the words of Max Weber, representing Occidental rationality as an alliance of bureaucracy, law, capitalism and technological progress. With his notion of the innovative entrepreneur as 'creative destroyer', Schumpeter added a heroic component to the social institution 'entrepreneurship', providing entrepreneurial practice with social legitimation and bundling up a pool of positive qualities entrepreneurial identification could draw from (Lindner 1999, p. 29).
- However, the entrepreneur acquires his status as a 'consequence of modernity' only in his functional dimension as risk-taking organizer and manager of inputs, selling goods/services in competitive markets as a means to achieve entrepreneurial ends. In this way the figure of the entrepreneur combines three of the central institutions of modernity (industrialism, capitalism, surveillance), which have been lifted out of national territorial contexts in a globalizing modernity (Giddens 1990, pp. 55–78).
- An indispensable precondition of current globalization processes, with regard to both the symbolic and the functional dimension of entrepreneurship, is the global 'triumph' of neoliberal ideas (Bierstecker 1995). Even without the direct involvement of the IMF or other multinational organizations the formally Socialist Republic of Syria has been undergoing a cautious process of economic liberalization and deregulation since the late 1980s (Perthes 1995, Hopfinger/Khadour 1998). Of crucial importance within this gradual transformation process was a 1991 investment law opening a large number of state-controlled monopolistic sectors for private capital. Private investors were successfully attracted by granting them numerous privileges (i.e. tax breaks, tariff exemptions, free capital transfer). Within a period of only five years Syrian entrepreneurs planned the establishment of more than 500 new industrial plants and the creation of almost 50,000 new jobs. Over 80% of the total volume of investment of more than US$3 bn – a considerable sum for a country with a GDP of US$12 bn – was to be financed through the repatriation of foreign assets (Hopfinger/Boeckler 1996). The fact that only just over half of the planned projects were realized by the end of the 1990s notwithstanding, there is an unmistakable tendency towards a withdrawal of the public and a simultaneous strengthening of the private sector. I do not want to engage in a discussion of whether the liberalization measures follow a macroeconomic rationale or whether they are only typical selective stabilization policies of an authoritarian regime in times of economic crisis (Heydemann 1992). It should suffice to mention that the accumulation crisis within the public sector, which Syria has faced since the 80s, led to a growing functional dependency of the economy on private sector participation.

(3) Far from only being linked to a legislative change of economic policies, economic liberalization, deregulation and systemic transformation are deeply implicated in a fundamental rearticulation of structures of signification, legitimization and domination, to use Giddens' words. Discredited until recently as bourgeois capitalists and marginalized politically by the state, entrepreneurs are now playing an active role in social processes. In their efforts to enforce and strengthen their functional power, for instance with the formation of intermediary interest groups, they actively appropriate the symbolic dimension of the figure of the Occidental entrepreneur and introduce it into public discourse. Being well equipped with economic and social capital, the new entrepreneurial class is relatively successful with these discursive manoeuvres. It is precisely this combination of functional necessity in the context of economic liberalization and symbolic power underlying the negotiation of new social norms that renders the entrepreneur so empirically interesting for an analysis of diacritical practices. From a perspective which conceptualizes organization with Giddens as "reflexive structuration" (Ortmann/Sydow/Windeler 1997), i.e. the reflexive structuring of rule-resource complexes in and through practice, and economic businesses as organizations which are above all characterized by the reproduction of allocative resources, the entrepreneur becomes the central figure in a context such as Syria. This does not endow entrepreneurs with the demiurgic capacity to shape organizations according to their intention. However, given the dominance of small firms and equity finance in the Syrian economy, owner-entrepreneurs are able to mobilize unevenly distributed power to reach specific arrangements of corporate rule-resource complexes: production targets, intra-firm hierarchies, codes of conduct, capital-labour relations etc. Furnishing the material design of company grounds and determining plant locations as well as production processes, they control and delimit workers' movements in time and space.

4.2 Syrian entrepreneurs and plurilocal places

4.2.1 Sociospheres and location

"Nous ne vivons pas à l'intérieur d'un vide (...), nous vivons à l'intérieur d'un ensemble de relations qui définissent des emplacements" (Foucault 1984, p.47). Accordingly, business enterprises may be conceptualized as ensembles of relations, with entrepreneurs playing a decisive role in the shaping of these placements. As such, the industrial plant is not placed into empty space. Rather, placement is an open process occurring through the knowing and intentional framing of unlimited relations. This is why Foucault proposes describing practices of placing by searching for the defining set of relations. Although Foucault himself relegated this project to the sidelines, preferring the solution of larger heterotopian problems, I would like to expand on this and look for some relational placings of Syrian business firms. From a perspective of Entterritorialisierung, and following Appadurai (1990) and Albrow (1997), these relational practices of placing may be divided into six imaginary/imagined landscapes of global cultural flows: the *ethnoscape* of mobile persons, the *ideoscape* of mobile ideas, world-views and ideologies, the *financescape* of mobile capital, the *technoscape* of mobile goods, machines and technology, the *mediascape* of mobile images and information and the *socioscape* (Albrow 1997) of subjectively constructed social relations. Through the structuration of business enterprises the diacritical practice of Syrian entrepreneurs bundles some of these translocal landscapes, landscapes which overlap and intersect but are in principal non-isomorphic. With Albrow (1997, pp. 51–52) one could read the business enterprise as a "sociosphere", a heterogeneous pattern of multiple symbolic systems, loosely held together by entrepreneurial efforts to standardize by cutting off the multiplicity of interwoven relations. Let me give you a brief example:

Just like many other Syrian entrepreneurs, Abu Zafer spent large parts of his life abroad. During the 1970s he studied economics in the US, before working as a contractor for infrastructure projects in various Gulf countries. When he finally returned to Aleppo in order to establish a pharmaceutical company in the early 1990s, he not only brought American instruments and machinery – surmounting administrative obstacles with the help of an "old friend". He also carried – to use his own words – "Japan to Aleppo". While it would be customary for Syrian entrepreneurs to treat "their employees like slaves, like animals", he tried to organize his plant in a way that assigns extensive responsibility to each member of the firm: "Everybody has to regard himself as part of a common whole". This is difficult to achieve, he concedes. On the one hand, he is expected to make the decisions: "In Syria, business is a one man show". On the other hand, he sees employees as being deeply loyal to their respective families. Accordingly, he has no grasp on the whole person, the company's influence ending at the walls of the plant. These obstacles notwithstanding, he considers himself on the right track.

Abu Zafer has never been in Japan. His imagination of 'Japan' has been filtered and broken in multiple ways through a translocal biography (e.g. as a student at a US business school, as consultant for Japanese companies in the Gulf). And correspondingly, he sees 'Syria' itself through glasses tinted by his experiences as 'migratory entrepreneur'. Abu Zafer's 'company sociosphere' charts the normative-significative space for the negotiation of meanings, but the renegotiation of norms, values and rules occurs to a large extent through the disjunctures of different 'scapes' rather than within the scapes themselves: for instance, through the gap which opens between the ideoscape of 'Japanese organizational paradigms' and the socioscape of 'family-bound solidarity'; or between the socially mediated technoscape of 'US machinery' and the ideoscape of 'anti-American sentiments' within the Syrian public sector (and anti-Syrian resentment in the US administration).

To Albrow (1997, p. 52) any person's place in a locality represents a point where his/her "sociosphere literally touches the earth"; accordingly, the sociospheres of newly built factories touch the earth at their location. In Syria the 1990s marked the beginning of the establishment of private industrial plants in the wider hinterland of larger cities. This not only changed the daily time-space paths of the workers in profound ways (bus transport, fixed departure times, etc.), it also fundamentally reconfigured social relations. The company forms a stage for the encounter of workers from rural and urban areas, actors with conflicting ideoscapes and diverging identity positions (proud farmers vs. decadent city dwellers and modern city dwellers vs. traditional farmers, respectively). If one understands the business enterprise as being interwoven with a local context through the located site of the firm, it follows that the specificity of 'placed locality' does not derive from an internal historical authenticity but from the interrelations of different sociospheres of different firms, producing plurilocality and transculturality, and at the same time being produced under these conditions.

4.2.2 Walls and rules

Bourdieu (1993, p. 163) speaks of "les sourdes injonctions et les rappels à l'ordre silencieux", silent 'calls to order' of material space. Social orders are imperceptibly incorporated into the bodies moving through place. Interpretative analyses of organizational cultures have demonstrated that the places of business enterprises are furnished with artefacts and accessories constantly uttering the silent rules of symbolic order. In the context of the Syrian industrial plant a particularly conspicuous substantialization of its social space concerns the surrounding brick wall. Day by day, bodies literally move through the tall barrier that demarcates the territorial limits of almost all plants founded in the 1990s. The methodological stance of Entterritorialisierung allows us to put an irritating question mark behind this apparently familiar and plausible material symbol. Interviewees are unable to provide a functional explanation for the wall: it protects neither against theft nor against espionage, piracy or government control. Bourdieu regards "spatial consumption" as an excellent device with which to demonstrate power. What precisely is the symbolic power expressed by the 'walled business enterprise'?

A stroll through the alleys of any of the new industrial zones north of Aleppo, accompanied by insurmountable, cold and distancing brick fences, inevitably evokes associations with ideas about the 'privacy' of the traditional 'Oriental city' (Wirth 1991). Should the uninviting exterior walls not be regarded as the architectonic reiteration of the typical, family-based socio-spatial organization of 'Oriental societies'? Such an interpretation may be supported with reference to the dominant role of the family as institution in economic affairs (Escher 1991, Lindner 1999). However, there might be a different approach: walls are borders. And just like any border they cut off connections and at the same time refer in and through the practice of separation to the underlying connectivity. Rather than starting the interpretation with the family, it may therefore be better to start with the fundamental connectivity of all (male) members of the society, as it is constructed by patrilinear kinship. From early on Western observers have classified this all-encompassing organizing principle of a (tribal) Orient with Durkheim's concept of 'segmentation'. Simply put, the elaborated 'segmentary lineage theory' encompasses the following elements: by patrilineal descent and the so-called 'preference marriage rule' for father's brother's daughter (bint 'amm) the virtual single universe in which everybody is akin with the same single ancestor is divided into ever smaller social units (tribal associations, tribes, clans, extended families, families). The borders between these social units are drawn by active internal solidarity and external 'complementary opposition'. The individual segments are endowed with equal rights, a pattern leading to the absence of stable political-organizational leadership. Traditionally, academic discussions (see for a summary Eickelman 1997, pp. 123–173) revolved around the question whether this reasoning provided an adequate description of Oriental societies or whether it is nothing but a concept invented by Western observers, more or less irrelevant for everyday life. What both positions fail to acknowledge is that the segmentary ideology was again and again reproduced discursively, while rarely being obeyed in social practice (Kraus 1995). Based on his own empirical work and including the spatiality of these social processes Salzman (1978) concluded the following: the ideology of segmentary lineage acquires practical relevance at times of territorial uncertainty. During periods of territorial stability, however, social practices deviate from these rules, bringing about new forms of social relations.

In this context, the 'walled business enterprise' represents exactly the opposite of the simple continuation of socio-spatial organization based on family ties and kinship. From the perspective of a given entrepreneur, the symbolic production of territorial stability is linked to the formulation of routines, scripts and norms for an 'organizational inside', which conflicts with the dominant norms of a 'societal outside': in this way Abu Zafer is able to bring 'Japan' to Aleppo, binding it to the physical-material 'stage' of his business enterprise. It is in this way that primordial ties are getting disentangled in order to negotiate the institution family at a different spatio-temporal scale. The following statement made by a successful textile entrepreneur serves as an example of the contractual reconstruction of the family behind the company walls:

> "For larger projects you need business partners. But partnership is a difficult thing here (...) probably because it is too easy to find a partner. You

know, there is a high amount of trust. People trust each other, it's almost compulsory. They come together, discuss shortly, 'this is your job, this is mine and the gain we share, let's do it', no time for details, and, because asking for a written contract would signal distrust, there is no written contract at all. But as soon as they face the first difficulties, they will argue and split up (...). We are different. We have detailed contracts even within the family. All responsibilities and rights, shares of capital and profits are fixed under witnesses. Even between me, my two brothers, my father and my mother. Besides, this appears in our computers as well (...)."

The symbolic structuration of the business enterprise occurs under translocal conditions of Entterritorialisierung. But the diacritical practice of Syrian entrepreneurs is at the same time linked to a material territorialization. Operating with socio-spatial inscriptions of specific qualities, it separates the organizational inside (reflexive structuration, capitalist industrialism) from a societal outside (primordial ties, traditionality). The simultaneity of separation and connection (purification and translation) in turn produces a local context which becomes meaningful only in its non-territorial 'Welt-Bindung'.

From the perspective of 'Entterritorialisierung' the organizational inside-outside differentiation is transformed into internal differences of an all-encompassing global inside. Therefore the rational modernity is not only produced by the separation of business and family as represented in Western discourse. The 'real differences' of Oriental modernity additionally emerge in and through the reflexive and contractual reinvention of the family within the boundaries of the walled business enterprise. Needless to say, family is just one dimension of the relational renegotiation of society through diacritical practice; there is a multitude of others, such as religion, symbolic value of commodities and so on.

4.2.3 Mediation and 'imaginative geographies'

The separating wall constitutes a material symbol for the local embeddedness of the translocally structured business enterprise itself. But it is only through the stabilized (transterritorial) territoriality that the business firm establishes new mediating connections to the non-organizational exterior. It is possible, for instance, to re-open the wall through a mosque in the plant's corner, accessible both from the inside and the outside, to cover the building's frontage with a prestigious design, or to raise the building's façade far above the wall, decorate it with the company's name, so that the new self-confidence of private capital in Syria becomes visible from a distance. Although transporting markedly different value systems, the material signs nevertheless converge around the intention to socially legitimize private economic success. But it is not only with material artefacts that Syrian entrepreneurs make powerful offers to resignify value hierarchies in their favour. Notwithstanding the fact that authoritarian regimes such as that in Syria normally prevent the organization of autonomous interest groups, entrepreneurs participate in public discourse and try to establish intermediary organizations at the inter-firm level. The following quote illustrates this.

Khalil Nour is a highly esteemed businessman from Aleppo boasting long-term and intensive contacts abroad who has been a leading partner of the country's largest company for the production of synthetic materials for 25 years. He has been involved in intensive efforts to establish associational structures in various business sectors for some years. In so doing, Khalil appropriates the symbolic dimension of the Occidental entrepreneur, stabilizing his own contradictory present through a discursive regionalization of the world with the image of the 'German industrialist', not unlike Abu Zafer's imaginary transfer of 'Japan' to Syria:

"In Germany industrialists are well respected and honored, their effort is acknowledged, here we are treated like thieves. This has to be changed, we have to change it (...) there, every industry has its own association (...) they work together and lobby for their industry (...) we have to become more active."

Essentialism is, as Gregory (1995, p. 457) argued, "a constitutive function of Orientalism itself." By essentializing the Orient, the Occidental discourse of Orientalism essentialized the Occident itself. Khalil's imaginary geographical construction of the 'German industrialist' breaks with this exclusive Eurocentric positioning of Orientalism referring to the fundamental relational nature of the world. This is the precondition for reciprocal 'real differences', in so far as "globally effective contexts perceive themselves as contexts of other contexts" (Nassehi 1998, p. 156). In this way, Khalil Nour places a "contextual universalism" in between the failing classical modernist attitudes of universalistic universalism and universalistic relativism (Beck 1997, pp. 141–149): Relativism fails in a world which cannot be neatly partitioned into separate entities. Universalism fails, given that interrelated worlds have different qualities. There is no alternative to "Einmischen, Eingemischt-Werden, Mit- und Gegenmischen" – to mutual and interrelated interferences – between the worlds (ibid., p. 145). Following from this, one might grant Syrian entrepreneurs the right to be involved with 'German entrepreneurs' with the construction of their own reciprocal 'imaginative geographies'. This could be a building block for a 'Kulturgeographie' (not only) of the Orient. The ability to construct regions and worlds is itself a global phenomenon. Appadurai (2000, p. 8) dropped en passant the term "recursive refraction", a well-chosen expression to denote the simple idea that all regions produce their own regions.

5. Orientalized geographies?

But what's left for the Orient?

"The Orient has a special character which is a blend of many religious beliefs, habits and traditions. Among those positive features or values are generosity, a community ethic and many others. Careful study of the characteristics of Oriental societies, Syria being one of these, elicit various positives, which if applied in industry together with high technology, will certainly yield distinguished results – even alongside those of the industrialized West" (Seiff 1998, p. 61).

Riad Seiff, who advanced this position during a conference in Germany, owns a textile plant in the vicinity of Damascus; with about 1,000 workers it is the biggest in the industry.[4] The plant supplies the Syrian market with 'German' sporting goods as a licensee. Since 1994 Seiff has been one of the few private sector representatives in the Syrian parliament. He sits amongst 'independent members' and uses the parliament and the public media as platforms to forcefully voice his demands for institutional reforms within Syrian bureaucracy and for an acceleration of stalled economic-political liberalization. If one interpreted Riad Seiff's statement with Edward Said's critique of Orientalism, one would have to regard Riad as a victim of hegemonic Orientalism, a victim who turns the discursive practice of "orientalizing the Orient" (Said 1995, p. 49) against himself. Considering that Riad Seiff jealously frames his cultural narrative of Syria's development prospects against the economic success of countries like Taiwan and South Korea and acknowledging that he links the positively-coded discussion about 'Asian values' to

4 When I wrote this paper, Riad Seiff secretly hoped to be appointed Minister of Industry after one of the next reshuffles of the Syrian cabinet, despite his publicly voiced criticism of economic policies. But the story had a different ending. As a symbolic support Riad was given the human rights award of the City of Weimar in 2003. What has happened? Hafez al-Assad's death in June 2000 initiated a short 'Damascene Spring', involving the organization of so-called diwaniyyas (political salons), that is, proto-party associations which promoted political reform. Being a leading figure of this movement at the beginning of 2001, Riad Seiff was first warned, then arrested and finally given a five-year prison sentence for 'threatening the Syrian constitution.'

his own situation, however, one could – with and against Said – give the recursive spiral a further twist: 'Oriental orientalizing the (orientalized) Oriental.'

This could also provide the answer to the questions about the nature of the 'Orient'. Is the 'Orient' not, amongst other things, the very product of the diacritical practice of Syrian entrepreneurs (or better: 'Oriental' entrepreneurs)? A practice which ties the worlds together, territorializes places and constructs local contexts, and, in this way, contributes particular stories to a global narrative of multiple modernities? Charting the ambivalent and fluid effects of diacritical practices from the vantage point of 'Entteritorialisierung' might be a worthwhile task for 'Oriental Geographies' interested in 'real differences' beyond the era of cultural realism in a global age.

6 References

Abu-Lughod, L. (1991): Writing against culture. R. G. Fox (ed.): Recapturing Anthropology. Santa Fe: School of American Research Press, pp. 137–162.

Agnew, J.A. (1994): The territorial trap: the geographical assumption of international relations theory. Review of International Political Economy 1, pp. 53–80.

Albrow, M. (1996): The Global Age: State and Society Beyond Modernity. Cambridge: Polity Press.

Albrow, M. (1997): Travelling beyond local cultures: socioscapes in a global city. J. Eade. (ed.): Living the Global City: Globalization as Local Process. London: Routledge, pp. 37–55.

Appadurai, A. (1990): Disjuncture and difference in the global cultural economy. Theory, Culture & Society 7, pp. 295–310.

Appadurai, A. (2000): Grassroots globalization and the research imagination. Public Culture 12, pp. 1–19.

Banse, E. (1926): Abendland und Morgenland: Landschaft, Rasse, Kultur zweier Welten (Occident and Orient: Landscape, Race and Culture of Two Worlds). Braunschweig: Westermann.

Barth, F. (1969): Ethnic Groups and Boundaries: The Social Organization of Cultural Difference. Boston: Little, Brown.

Bauman, Z. (1997): Postmodernity and its Discontents. Cambridge: Polity Press.

Beblawi, H. (1987): The rentier state in the Arab world. H. Beblawi and G. Luciani (ed.): The Rentier State: Nation, State and Integration in the Arab World. London: Croom Helm, pp. 49–62.

Beck, U. (1986): Risikogesellschaft. Auf dem Weg in eine andere Moderne. Frankfurt/M.: Suhrkamp. – (1992) Risk Society: Towards a New Modernity. London: Sage.

Beck, U. (1997): Was ist Globalisierung? Irrtümer des Gobalismus – Antworten auf Globalisierung (What is Globalization? Fallacies of Globalism – Responses to Globalization). Frankfurt/M.: Suhrkamp.

Bhabha, H. (1994): The Location of Culture. London: Routledge.

Bierstecker, T.J. (1995): The 'triumph' of liberal economic ideas in the developing world. B. Stallings (ed.): Global Change, Regional Response. Cambridge: Cambridge University Press, pp. 174–196.

Boas, F. (1927): Primitive Art. – Institut for sammenlingnende Kulturforskning 8. Oslo: H. Aschehoug & Co.

Bourdieu, P. (1979): La distinction: critique sociale du jugement. Paris: Les Èditions de Minuit. – (1984). Distinction: A Social Critique of the Judgement of Taste. Cambridge, MA: Harvard University Press.

Bourdieu, P. (1993): Effets de lieu. P. Bourdieu (ed.): La misère du monde. Paris: Édition du Seuil, pp. 159–167. – (1999). Site Effects. P. Bourdieu (ed.). The Weight of the World. Social Suffering in Contemporary Society. Oxford: Polity Press, pp. 23–36.

Bourdieu, P. and Wacquant, L. (1999): On the cunning of imperialist reason. Theory, Culture and Society 16, pp. 41–58.

Clifford, J. (1997): Routes: Travel and Translation in the Late Twentieth Century. Cambridge, MA: Harvard University Press.

Deleuze, G. and F. Guattari (1986): Nomadology: The War Machine. New York: Semiotext(e).

Derrida, J. (1972): Marges. De la Philosophie. Paris. – (1982). Différance. In: J. Derrida (ed.). Margins of Philosophy. Chicago: University of Chicago Press, pp. 3–27.

Derrida, J. (1988): Limited Inc. Evanston: Northwestern University Press.

Eickelman, D.F. (1997): The Middle East and Central Asia: An Anthropological Approach. Upper Saddle River, NJ: Prentice Hall.

Elias, N. (1976 [1939]): Über den Prozeß der Zivilisation. Frankfurt/M.: Suhrkamp. – (1994) The Civilizing Process. Oxford: Blackwell.

Elias, N. (1987 [1939]): Die Gesellschaft der Individuen. Frankfurt/M.: Suhrkamp. – (1991) The Society of Individuals. Oxford: Blackwell.

Elias, N. (1996 [1970]): Was ist Soziologie? Weinheim: Juventa. – (1978). What is society? London: Hutchinson.

Escher, A. (1991): Sozialgeographische Aspekte raumprägender Entwicklungsprozesse in Berggebieten der Arabischen Republik Syrien (Social Geography of Spatial Development in Mountain Areas of Syria). – Erlanger Geographische Arbeiten 20. Erlangen: Fränkische Geographische Gesellschaft.

Fisch, J. (1992): Zivilisation, Kultur. O. Brunner, W. Conze and R. Koselleck (eds.): Geschichtliche Grundbegriffe: Historisches Lexikon zur politisch-sozialen Sprache in Deutschland, Bd. 7. (Fundamental Concepts: Historical Encyclopaedia of Socio-political Language in Germany, Vol. 7). Stuttgart: Klett-Cotta, pp. 679–774.

Foucault, M. (1984 [1967]): Des espaces autres. Architecture, Mouvement, Continuité 5, pp. 46–49. – (1986). Of other spaces. diacritics, spring 1986, pp. 22–27.

Friedman, J. (1994): Cultural Identity and Global Process. London: Sage.

Friedman, J. (1999): The hybridization of roots and the abhorrence of the bush. M. Featherstone and S. Lash (eds.): Spaces of Culture. London: Sage, pp. 230–256.

Geertz, C. (1973): Thick description: toward an interpretive theory of culture. C. Geertz (ed.): The Interpretations of Cultures. London: Fontana Press, pp. 3–30.

Giddens, A. (1979): Central Problems in Social Theory: Action, Structure and Contradiction in Social Analysis. Berkeley: University of California Press.

Giddens, A. (1984): The Constitution of Society: Outline of the Theory of Structuration. Berkeley: University of California Press.

Giddens, A. (1987): Structuralism, poststructuralism and the production of culture. A. Giddens and J. Turner (eds.): Social theory today. Cambridge: Polity Press, pp. 195–223.

Giddens, A. (1990): The Consequences of Modernity. Cambridge: Polity Press.

Gregory, D. (1995): Imaginative geographies. Progress in Human Geography 19, pp. 447–485.

Hall, S. (1990): Cultural identity and diaspora. J. Rutherford (ed.): Identity: Community, Culture, Difference. London: Lawrence and Wishart, pp. 222–237.

Hall, S. (1992): The question of cultural identity. S. Hall, D. Held and T. McGrew (eds.): Modernity and its Futures. Cambridge: Polity Press, pp. 273–316.

Hall, S. (1996): When was 'the postcolonial'? Thinking at the limit. I. Chambers and L. Curti (eds.): The Postcolonial Question: Common Skies, Divided Horizons. London: Routledge, pp. 242–260.

Hansen, K.P. (1992): Die Mentalität des Erwerbs. Erfolgsphilosophien amerikanischer Unternehmer (Mentality of 'Earning': American Entrepreneurship and Philosophies of Success). Frankfurt/M.: Campus.

Heidegger, M. (1990 [1954]): Bauen – Wohnen – Denken (Building – Dwelling – Thinking). M. Heidegger (ed.): Vorträge und Aufsätze (Lectures and Essays). Pfullingen: Neske, pp. 139–156.

Heydemann, S. (1992): The political logic of economic rationality: selective stabilization in Syria. H. J. Barkey (ed.): The Politics of Economic Reform in the Middle East. New York: St. Martin's Press, pp. 11–39.

Hopfinger, H. and Boeckler, M. (1996): Step by step to an open economic system: Syria sets course for liberalization. British Journal of Middle Eastern Studies 23, pp. 183–202.

Hopfinger, H. and Khadour, R. (ed.) (1998): Wirtschaftsentwicklung und Investitionspolitik in Syrien/Economic Development and Investment Policies in Syria. – Eichstätter Geographische Arbeiten 10. München: Geobuch.

Huntington, S. (1996): The Clash of Civilizations. New York: Simon & Schuster.

Kopp, H. (1999): Recent German human geographical research on the Arab world. The Arab World Geographer 2, p. 259.

Kraus, W. (1995): Segmentierte und segmentäre Theorie: Strukturelle und kulturelle Grundlagen tribaler Identität im Vorderen Orient (Theory of segmentation and segmentary theory: structural and cultural foundations of tribal identity in the Middle East). Sociologus 45, pp. 1–25.

Latour, B. (1993). We have never been modern. Cambridge, MA: Harvard University Press.

Latour, B. (1998): From the world of science to the world of research? Science 280, pp. 208–209.

Law, J. (1992): Notes on the theory of the actor network: ordering, strategy and heterogeneity. Systems Practice 5, pp. 379–393.

Lindner, P. (1999): Räume und Regeln unternehmerischen Handelns: Industrieentwicklung in Palästina (Rules and Spaces of Entrepreneurial Agency: Industrial Development in Palestine). – Erdkundliches Wissen 126. Stuttgart: Franz Steiner.

Luhmann, N. (1997): Die Gesellschaft der Gesellschaft (The Society of Society). Frankfurt/M.: Suhrkamp.

Massey, D. (1999a): Imagining globalisation: power-geometries of time-space. D. Massey (ed.): Power-Geometries and the Politics of Space-Time: Hettner Lecture 1998. – Hettner Lectures 2. Heidelberg: University of Heidelberg, Department of Geography, pp. 9–23.

Massey, D. (1999b): Philosophy and politics of spatiality: some considerations. Geographische Zeitschrift 87, pp. 1–12.

Nassehi, A. (1998): Die 'Welt'-Fremdheit der Globalisierungsdebatte (The strange worlds of globalization discourse). Soziale Welt 49, pp. 151–166.

Ortiz, F. (1940): Contrapunteo cubano del tabaco y el azúcar. Havana: Jesus Montero. – (1995): Cuban counterpoint: tobacco and sugar, introduction by Bronislaw Malinowski. London: Duke University Press.

Ortmann, G., Sydow, J. and Windeler, A. (1997): Organisation als reflexive Strukturation (Organization as reflexive structuration). G. Ortmann, J. Sydow and K. Türk (ed.): Theorien der Organisation: Die Rückkehr der Gesellschaft (Theories of Organization: The Return of Society). Opladen: Westdeutscher Verlag, pp. 315–359.

Perthes, V. (1995): The Political Economy of Syria under Asad. London: I.B. Tauris.

Pratt, G. (1999): Geographies of identity and difference. D. Massey, J. Allen and P. Sarre (eds.): Human Geography Today. Cambridge: Polity Press, pp. 151–167.

Robertson, R. (1992): Globalization: Social Theory and Global Culture. London: Sage.

Said, E. (1995 [1979]): Orientalism: Western Conceptions of the Orient. New York: Penguin.

Salzman, P.C. (1978): Does complementary opposition exist? American Anthropologist 80, pp. 53–70.

Sayigh, Y.A. (1957): Toward a theory for entrepreneurship for the Arab East. Explorations in Entrepreneurial History 1, pp. 123–127.

Schatzki, T.R. (1996): Social Practices: A Wittgensteinian Approach to Human Activity and the Social. Cambridge: Cambridge University Press.

Schlüter, O. (1906): Die Ziele der Geographie des Menschen (The Aim of Human Geography). München: Oldenbourg.

Schumpeter, J. (1934 [1911]): Theorie der wirtschaftlichen Entwicklung: Eine Untersuchung über Unternehmergewinn, Kapital, Kredit, Zins und den Konjunkturzyklus. Berlin: Duncker & Humblot. – (1934): The Theory of Economic Development. Cambridge, MA: Harvard University Press.

Seiff, R. (1998): Private sector development: scenarios for the future. Perthes, V. (ed.): Scenarios for Syria: Socio-Economic and Political Choices. Baden-Baden: Nomos, pp. 59–70.

Smith, A.D. (1979): Nationalism in the Twentieth Century. Oxford: Oxford University Press.

Thrift, N. (1999): Steps to an ecology of place. D. Massey, J. Allen and P. Sarre (eds.): Human Geography Today. Cambridge: Polity Press, pp. 295–322.

Weber, M. (1988 [1920]): Gesammelte Aufsätze zur Religionssoziologie. Tübingen: J.C.B. Mohr (Paul Siebeck). – (2001): The Protestant Ethic and the Spirit of Capitalism. London: Routledge.

Welsch, W. (1999): Transculturality: The puzzling form of cultures today. Featherstone, M. and Lash, S. (ed.): Spaces of Culture. London: Sage, pp. 194–213.

Werlen, B. 1997. Sozialgeographie alltäglicher Regionalisierungen. Band 2: Globalisierung, Region und Regionalisierung (Social Geography of Everyday Regionalization. Vol 2: Globalization, Region and Regionalization). – Erdkundliches Wissen 119. Stuttgart: Franz Steiner.

Wirth, E. (1971): Syrien: Eine geographische Landeskunde (Regional Geography of Syria) Darmstadt: Wissenschaftliche Buchgesellschaft.

Wirth, E. (1991): Zur Konzeption der islamischen Stadt (Conceptualizing the Islamic city). Die

Welt des Islams 31, pp. 50–92.
Zaimoglu, F. (1998): Koppstoff: Kanakasprak vom Rande der Gesellschaft ('Koppstoff': Other voices from the Margins of Society) Hamburg: Rotbuch.
Zierhofer, W. (1999): Geographie der Hybriden. Erdkunde 53, pp. 1–13.

Address of the author: Marc Boeckler, Department of Geography, Catholic University of Eichstätt-Ingolstadt, Ostenstr. 26–28, 85072 Eichstätt, marc.boeckler@ku-eichstaett.de

First published in Geographische Zeitschrift vol. 87/1999 (3 + 4), pp. 178–193

Geographische Zeitschrift, Special issue 2004

Smallholder vulnerability in degraded areas

The political ecology of pioneer frontier processes in Brazil

MARTINA NEUBURGER, Tübingen

Summary

The study of the relationship between socio-economic and political structures and ecological processes is of growing significance in geographical development studies, especially in the field of political ecology. For this study, the concepts of vulnerability, fragility and criticality are used. In studies of rural areas, the so-called land manager represents the main focus of interest. This is because, as the actor at the local level, the land manager is at the interface between ecology and socio-economics. In this context, pioneer frontiers serve as refuge areas for the displaced groups. With their incorporation into national economic and social structures after a short period of time, however, this function is rapidly lost again.

In this paper, the significant factors and processes of development at the frontier are investigated using an example from the Brazilian Mid-West. Processes of ecological degradation in the pioneer frontiers of the Amazon region can be understood as the result of global, national and regional/local structures, as a result of which particularly vulnerable groups (in this case peasants) are displaced into ecologically fragile areas in the course of time. The survival-oriented exploitation of natural resources in these refuge areas is in accordance with their logic of action. This logic, which must be seen as a strategy of survival, is based on decisions made between vulnerability and constraints in everyday situations.

1. Introduction

The growing environmental problems in the Third World have been the cause of much recent political, social and scientific debate, in the course of which it has become increasingly apparent that ecological, socio-economic and political processes are closely interlinked. The ensuing general scientific analysis of man-environment relationships has shown that an unfavorable constellation of factors can lead rapidly to crisis. Crises in this context are to be understood as events which result in material, social or psychological damage, and which demand a response from those affected. The frequency and impact of crises vary considerably as regards their spatial, temporal and group specificity. The concept of vulnerablity, originating from hazard research, has been developed to provide a conceptual framework to describe and analyse the factors determining the extent to which an individual or group may be affected by such crises (Hewitt 1997, Bohle 1998).

The concept of vulnerability includes three dimensions: 1) *risk susceptibility* describes the likelihood of a certain population group suffering from the negative consequences of a crisis; 2) *resistance* is defined as the ability to respond to such a critical situation, and depends on the range of coping strategies available; 3) finally, *ability to regenerate* is the ability to recover after a crisis and to restore things to their original state.

The degree of vulnerability of an individual or group results from the interaction of these dimensions, and depends on the specific combination of socio-economic, political, cultural-psychological and ecological factors (Blaikie, Cannon & Wisner 1994). Changes in risk factors over time make it necessary to distinguish between everyday and acute vulnerability. Everday vulnerability refers to long-term, structural, system-inherent vulnerability. The individual or group affected resorts to everyday survival strategies. Acute vulnerability, however, can be traced to short term causes. They are for the most part outside the system and require additional coping strategies. However, the implementation of unsuitable short-term survival strategies to overcome an acute crisis may increase everyday vulnerability, and have negative long-term consequences. In particular, persons or groups who already have to use all available resources in coping with day-to-day life to maintain the status quo, run the risk of increasing vulnerability to new crises due to a lack of alternatives and the use of "wrong" strategies.

So-called land managers, acting as mediators between ecology and the social economy at a local level, are the focus of interest in studies of ecological degradation, which causes many crises in developing countries. Human exploitation of nature, and ecological changes interact mutually. Human actions are influenced by numerous cultural, socio-economic and political factors, while the response of the natural environment to human use depends on the nature of its system-inherent characteristics, i.e. sensitivity, resilience and fragility (Goudie 1994, Holling 1980).

The concept of criticality has been developed in an attempt to combine approaches from social sciences and natural sciences (Kasperson, Kasperson, Turner II, Dow/Meyer 1995). Here the degree of vulnerability of a group in a specific region is juxtaposed with the degradation of the ecosystem. The ability of the natural environment to respond and to regenerate (resilience) as well as that of society (response systems) are the most significant elements. Criticality in this context constitutes the state of greatest risk. This means that a high level of criticality has been reached when vulnerable groups in degraded areas with high ecological fragility have run out of options, so that a crisis threatens to result in catastrophe, in the absence of external intervention. Thus the extent to which environmental degradation has already progressed is crucial to the criticality of a region, as well as the consequences of this degradation for the local population's prosperity and quality of life, and the extent to which any options remain which allow for long-term human use of the region by applying management strategies (Manshard/Mäckel 1995). As regions are neither ecologically, socio-economically nor politically closed systems, interactions and influences at different scales and levels must also be taken into consideration.

Political-ecological analyses concentrate on the examination of these constellations of socio-structural factors and their correlation with processes of ecological degradation, and are growing in significance in geographical development studies. In Third World countries in particular, serious socio-political and economic disparities are caused by growing tendencies towards marginalisation and by increasing environmental problems (Blaikie/Brookfield 1987, Bryant/Bailey 1997).

Pioneer frontier regions are particularly suited to the empirical analysis of the correlation between social structures and land degradation described in this essay, as these peripheral areas serve as areas of retreat for displaced groups (Peet/Watts 1996, Painter/Durham 1995). In the course of the incorporation of frontier regions into national economic and social structures, the disparities of the national centre are reproduced in the periphery, so that the pioneer frontier loses its function as a space for survival (Schmink/Wood 1986, 1992, Coy/Lücker 1993).

These processes are typical of frontier development and its relocation in a large nation such as Brazil, where its size, economic potential and the supposedly deserted interior have attracted many different actors since colonial times. Over the centuries, changing social, political, and economic factors have determined the spatially and temporally differentiated types of land exploitation in Brazil. Smallholder groups have always been pushed into locations which are marginal in terms of ecology and socio-economics, especially frontier regions, where they have been forced to fight for their survival.

The following paper analyses the correlation of pioneer frontier development and smallholder vulnerability based on a case study from the Brazilian Mid-West, beginning with a general introduction to the development of pioneer frontiers in Brazil. The political-ecological approach requires that three different analytical dimensions be taken into account (Geist 1992):

- The factors influencing pioneer frontier development are examined at different scales by comparing international, national, regional, and local frontier processes in a vertical analysis.
- In order to do justice to the complexity of frontier processes and to establish the correlation of vulnerability and ecological degradation, a horizontal-sectoral analysis examines the relationships between socio-economic, political and ecological determinants.
- Finally, historical development will be analysed.

2. Pioneer frontier development, ecological degradation and peasants in Brazil

The development of Brazil's present-day territory since the arrival of the Portuguese in 1500 shows the relationship between social vulnerability and ecological degradation in a historical perspective. It becomes clear that especially vulnerable groups are always pushed into ecologically fragile or spatially marginalised areas, where their options are extremely limited. If these retreats become integrated into economic and social structures, vulnerable groups may be displaced yet again. Processes of land acquisition and rights of disposal, determined by political-economic power structures, result in modifications of the ecological setting.

Before the arrival of European colonists, Brazil was inhabited by a multitude of indigenous groups who had close relations with each other and made use of the natural environment on the basis of their knowledge and experience of several millennia (see Fig. 1). The extensive, mostly extractive, practices of utilisation which they employed meant that the effects on the ecosystem took the form of gradual change rather than lasting damage (Guidon 1992, Roosevelt 1992).

As colonisation began, indigenous groups were driven far into the Brazilian interior, mostly into ecologically unsuitable areas. Portugal, a nation integrated into the political and economic power structures of Europe, took charge of the country. The exploitation of the colony's resources was intended to strengthen the mother country's position of power (Martins 1997). As no precious metals were found, the Portuguese concentrated on agriculture to make the best use of their overseas territories. After the rapid depletion of the forests, the colonial power granted colonists large areas of land with the intention of introducing large scale sugar cane cultivation. Indigenous tribal land tenure practices and laws were entirely ignored. The vast amounts of land available caused it to be perceived as a disposable resource and the sugar barons saw no necessity to strive for long-term preservation of the natural soil fertility. Ruthless profit-oriented exploitation of the land initiated processes of rapid ecological degradation (Dean 1996).

In the 18th century, indigenous groups were pushed further into the Brazilian interior after extensive gold and diamond deposits were found in their settlement areas. Mining and the associated population growth had far-reaching ecological consequences. In addition to devastation of the surface-mining areas, the

	region	predominant interests	function of the *frontier*	smallholder groups	land rights	spatial system
pre-Columbian	all of Brazil	local, regional	subsistence, survival strategies	indigenous groups with rural economic system	collective (tribal rights)	
colonial times (1500 until end of 17 th century)	coastal regions	global	accumulation of capital, maximisation of profit	indigenous population in Jesuit settlement	feudal (colonial rights)	
colonial times (18 th century)	hinterland insular around gold deposits	global	accumulation of capital, maximisation of profit, safeguarding of land	free peasants near the cities (cultivation of staple foods)	feudal (colonial rights)	
empire 1st republic (19 th until start of 20 th century)	southeastern hinterland, South Brazil, rivers of Amazonia	global, national	accumulation of capital, maximisation of profit	free peasants in the South, dependent peasants in the Southeast and the Northeast	private (constitu-tional law)	
Estado Novo **military dictatorship (1930s to 1980s)**	South Brasil, Amazonia	global, national	accumulation of capital, safeguarding of land, survival strategies	colonists, occupants of land, leaseholders	private (constitu-tional law)	
democracy (since mid-1980s)	Amazonia standstill of the *frontier*?	global, local	survival strategies, nature preservation, sustainability	colonists, occupants of land, leaseholders	private, collective, corporational (constitutional law)	

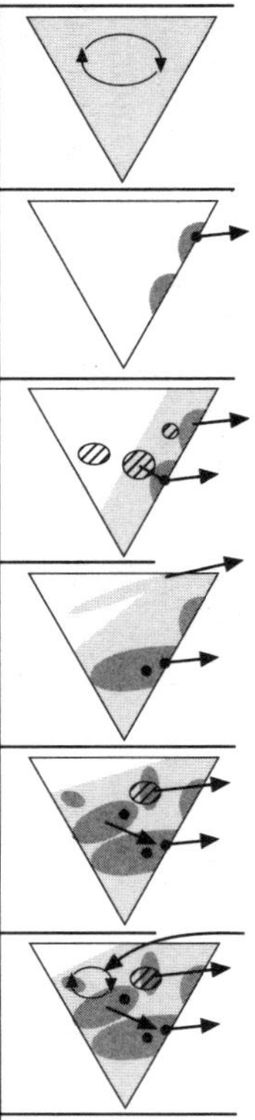

area model of Brazil

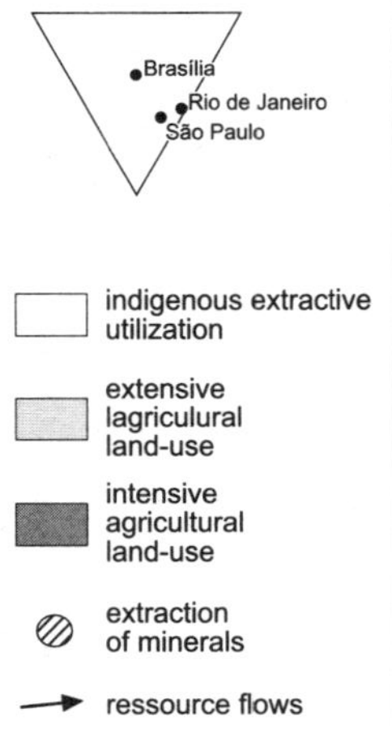

Neuburger 1999

Fig. 1: Pioneer frontier development in Brazil

forests in the vicinity fell victim to the enormous demand for timber. Staple food production expanded. The colonial power also supported the expansion of livestock farming, primarily in order to secure the mineral deposits far beyond the *Linha de Tordesillas*[1] against Spanish territorial claims.

While a smallholder class was virtually non-existent in the slave-holder society of the 16th and 17th centuries, in the 18th century this class expanded considerably as it became responsible for supplying the growing cities with food (Woortmann & Woortmann 1997). The peasants, as a newly-developing social group in Brazil which exhibited a high degree of vulnerability similar to the indigenous population, developed a conservationist use of nature in a system combining crop rotation and shifting cultivation. However, some degree of ecological damage was inevitable as the hilly hinterland of the mineral deposits was extremely susceptible to erosion (Dean 1996).

Although a national government assumed control of the country when Brazil became independent in 1822, its dependence on Europe did not cease to weigh down the former colony. England insisted that the new state bear part of Portugal's high debt. This colonial heritage inevitably led to an increased exploitation of natural resources. Now the economic and political elite did not only have to supply their own needs for profit and luxury goods, but also had to improve public finances by means of exports.

As mining had become less important, coffee was the only export item which guaranteed adequate profit margins in the 19th century. The advance of coffee was associated with an enormous expansion of the agriculturally exploited area in the southeast of the country (Kohlhepp 1990). There were no obstacles at all, as the conversion of tenure into private proprietary rights, and the appropriation of hitherto fallow land and its transformation into government property made virtually no difference to the practice of arbitrary seizure of land and ruthless exploitation of the natural resources on the part of the coffee barons. The weak, young state had no control mechanisms (Dean 1996, Margolis 1977). Once again, indigenous groups and smallholders were the victims of this development. The large landowners drew on them as slaves and cheap labour.

With the abolition of slavery in 1888, a great shortage of agricultural labourers set in, so that the Brazilian government increasingly recruited European immigrants. The immigrants, who mostly came from Italy and Spain, worked as labourers and tenants for the coffee barons in order to pay for their passage retrospectively. Also, the majority of the former slaves continued to work as labourers on coffee or sugar cane plantations. Thus, a large class of smallholders came into being, characterised by extreme dependence.

In order to provide the growing urban market centres with staple foods, mainly German, Southern and Eastern European smallholders were settled in the south of the country. The usual practice of settling the immigrants in planned colonisation projects meant that they generally possessed secure titles to land. Because of their strong entrepreneurial spirit and good connections to the market centres, they were able to generate some capital. This capital has recently formed the basis of modernisation in the region (Lücker 1986).

While the coffee frontier led to marked ecological degradation in southeastern Brazil, the crop rotation practiced by the well-structured and thus non-vulnerable smallholders in the south conserved the comparably stable ecosystem, and the rubber boom in the Amazon region had hardly any negative effect on the natural environment. As the forest was essential to the rubber tree's (*Hevea brasiliensis*) ability to reproduce and also for the long-term

1 The *Linha de Tordesillas* (Line of Tordesillas) was drawn on the basis of an arbitral award by Pope Alexander VI in 1494. It coincided with the 46th west meridian of longitude and separated the Spanish and Portuguese dominions in the New World.

realisation of profits, the extraction of rubber was a sparing use of resources (Kohlhepp 1987).

The 20th century was characterised by the opening-up and exploitation of the Amazon basin. As early as the 1930s, president Getúlio Vargas heralded this development with the catchphrase *Marcha para Oeste* – the westward march. Agricultural colonisation, first implemented in the 1960s and 1970s, involved the settlement of landless families from the critical rural areas of the northeast, and also from the south and southeast. These measures, which were supposed to render agrarian reform unnecessary, turned the Amazon region into the new retreat for vulnerable smallholder families (Kohlhepp 1979). Also, the function of the Amazon periphery as a *válvula de escape* or "safety valve" for the growing social tensions in the old settlement areas has been maintained to this day. At the same time, profit-oriented investors and members of the military embarked on strategies to secure economic and geo-strategic advantages in the development of the region.

This very mixed nature of the overlapping social requirements made of one region, inevitably led to violent conflicts (Coy & Neuburger 1999). Groups which used the periphery as survival areas, especially smallholders and *índios*, competed with profit-oriented actors such as the timber industry, owners of large estates and mining companies for space and resources. The resulting highly complex constellations with varied and ever-changing coalitions of interests, are typical of current processes and structures in the Amazon region (Martins 1990, Schmink/Wood 1986, 1992). This is also reflected in the growing ecological degradation of this region. Recently an attempt has been made to halt this development by applying sustainable development concepts at various scales (Hall 1997).

The analysis of pioneer frontier development in Brazil shows that disparate social and eocnomic structures and the unequal distribution of political power have been reproduced in frontier regions, due to their involvement in national and international processes since colonial times. The frontier serves as a retreat and survival space for vulnerable groups such as smallholders who, due to the expansion of profit-oriented utilisation practices, are pushed into ever more peripheral areas. The signs of serious ecological degradation are a consequence of this and correspond to the exploitation patterns of the respective groups in the individual areas.

3. The hinterland of Cáceres: Pioneer frontier development and ecological degradation

In the hinterland of Cáceres, situated in the southwest of the state of Mato Grosso, the relationships described above can be observed at a regional level. With regard to the processes described in connection with pioneer frontiers, the region represents a link between the development of the old settlement areas of Brazil and the colonisation of the Amazon periphery. The settlement projects of the 1950s and 1960s in this area were the hesitant beginnings of these state-run initiatives in development politics which were to lead to more extensive development measures in the Amazon region later on. In the following analysis of the regional historical processes, we will see which factors and interdependencies influence the spatial patterns of socio-economic and ecological structures in the context of the political-ecological approach. For this analysis, special attention has to be paid to the incorporation of smallholder groups into frontier development, as well as the changes in the degree of their vulnerability in the course of these processes.

Just as in almost all of Brazil's pioneer frontier areas, the planned increase in settlements in the Cáceres hinterland using external actors superseded a low-impact, extractive utilisation, the extraction of ipecac (*Cephaëlis ipecacuanha*), which in its turn had displaced the indigenous groups. The complex regional constellations of actors and associated planned

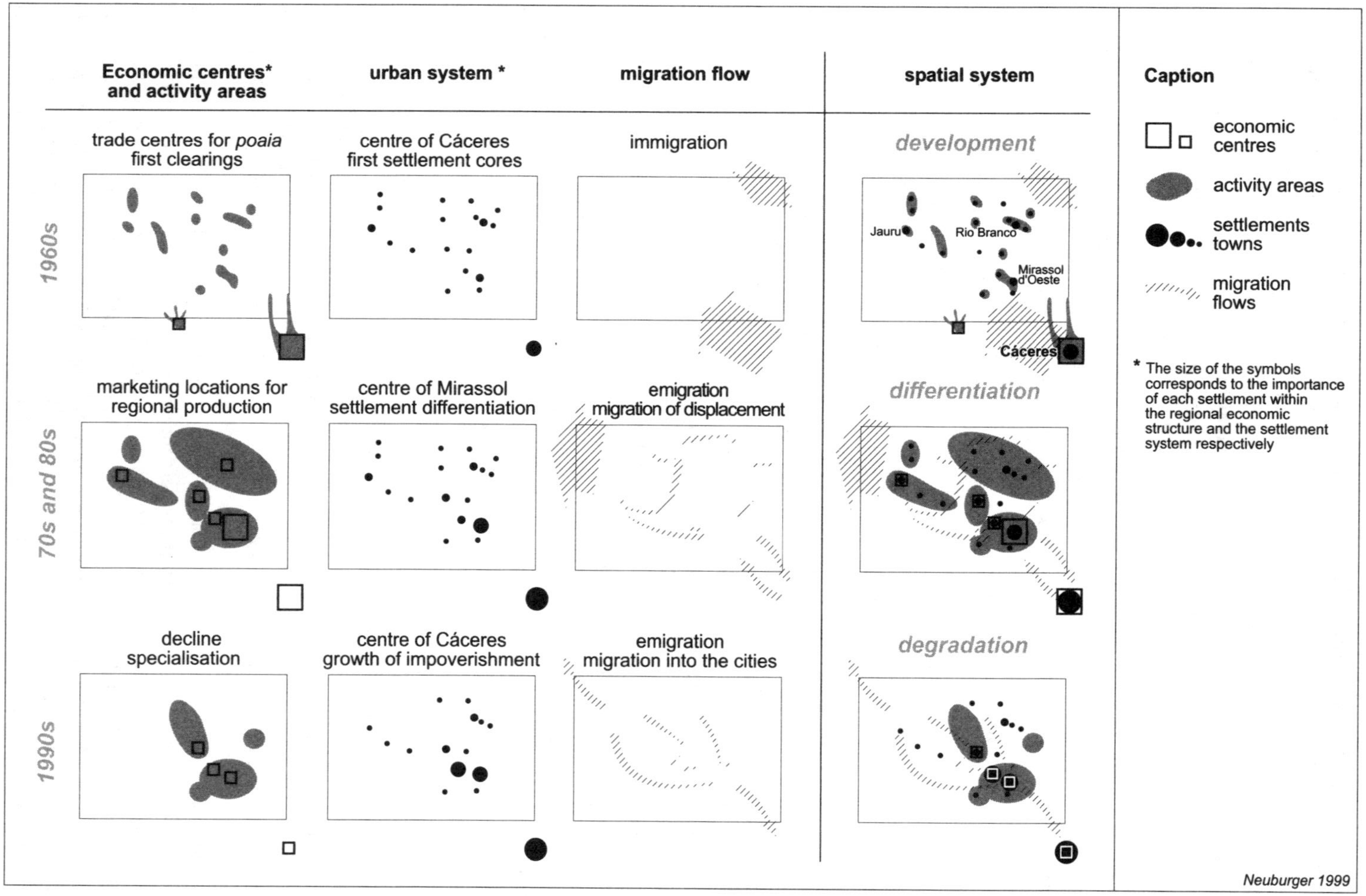

Fig. 2: Changes in the spatial system in the Cáceres hinterland

and spontaneous processes led to the emergence of locally and regionally differentiated socio-economic structures.

The first signs of disparate spatial structures emerged as early as the development phase of the 1960s (see Fig. 2). In the eastern and western border areas, state-run or semi-state colonisation projects dominated, whereby poor smallholders from Minas Gerais and Espírito Santo received land free of payment. In the south-central areas, private actors shaped the peasant colonisation, in some places medium- and large-scale companies were also involved, all of which were well funded by settlers from São Paulo, Goiás and Paraná.

This socio-spatial constellation of different degrees of vulnerability corresponded to the spatial patterns of ecological fragility in the region. The central and southern regions were ecologically favourable, and their better transport infrastructure meant that they also had other positive locational factors. The northern and border areas, however, where sandy soils and high relief energy dominated, were relatively unsuitable for agricultural use.

Thus, the foundations had been laid for disparate structures in the Cáceres hinterland, which were to be intensified during the differentiation phase. The 1970s were initially characterised by economic prosperity and high agricultural output, which was marketed all over Brazil via Cáceres. Internal exchange also became more important with the expansion of the road networks within the region. The city of Mirassol d'Oeste, where industrial, commercial and service functions as well as infrastructure facilities were concentrated, held a central position in the regional settlement system.

Within the first few years the particularly vulnerable smallholder farms in the ecologically less desirable areas suffered severe drops in staple food production due to rapid soil degradation. As a consequence, land concentration processes and then displacement set in. Better ecological preconditions meant that these economic and ecological crises did not occur in the private colonisation projects, which were dominated by coffee cultivation, until the 1980s. Here, tenants working on coffee plantations in particular had to migrate when coffee cultivation declined.

At the end of this phase, the majority of farmers gave up tillage and turned to cattle farming. While the medium- and large-scale farms for the most part switched to an intensified form of farming, fat stock and breeding cattle, the smaller farms dabbled in extremely precarious dairy cattle farming. Yet most of them failed. Increased inflaction and the stagnation of prices caused by national agrarian policy were responsible for this development. Displacement migration ensued, the main flow of which concentrated on the new pioneer frontier regions in northern Mato-Grosso and Rondônia. Many smallholders, however, also occupied fallow areas within the region leading to numerous violent conflicts over land.

The complex spatial system of the 1970s and the 1980s simplified with growing socio-economic and ecological degradation in the 1990s. With the rapid decline of the total population in rural areas, medium- and large-scale cattle farms established themselves in the region and left very little space for small-scale farms. Only the ecologically least favourable, most fragile areas were left for smallholder agriculture. The larger farms, however, intensified and specialised their production in the favourable areas in order to meet the challenges of a liberalised agricultural market in the 1990s (Remppis 1998). Only the relatively well-funded small-scale farms in the environs of Mirassol d'Oeste, which had grown coffee before, were able to find a niche in the regional market using innovative forms of production.

Even today, smallholder families continue to migrate from the almost depopulated rural areas into the regional cities, while migration to current pioneer frontiers has for the most part become unattractive as many have heard about former smallholders' failures. Squatter settlements in the cities, which started to de-

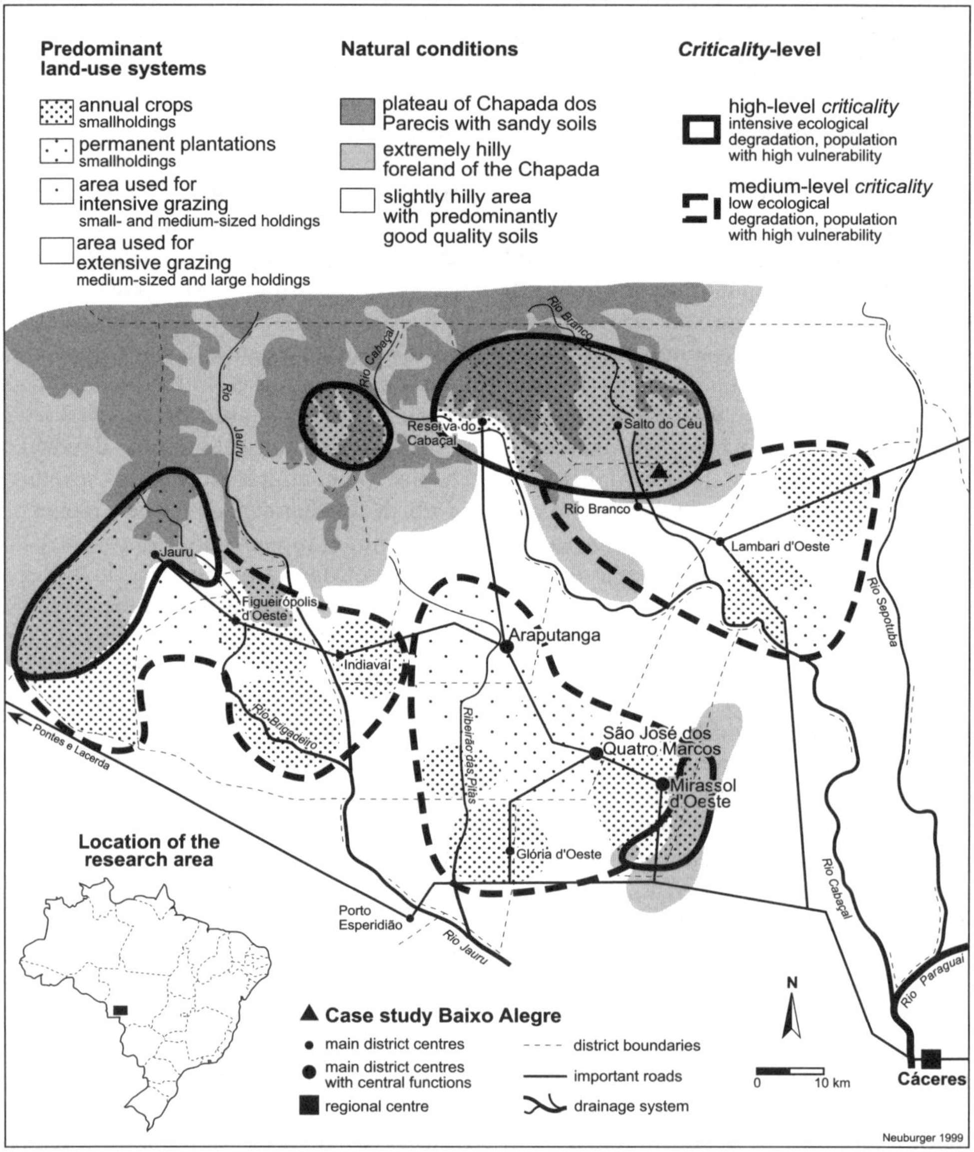

Fig. 3: Criticality in the Cáceres hinterland

velop in the 1980s, keep growing. Thus, social and ecological problems have shifted from rural areas into the cities.

The regionally and locally differentiated pioneer frontier development in the Cáceres hinterland have produced a socio-economic spatial pattern which largely corresponds to the basic ecological patterns (see Fig. 3). The displacement of smallholder groups is especially pronounced in the plateau regions of the *Chapada dos Parecis*, where marked natural fragility meets with high smallholder vulnerability. Today, only large-scale farms are to be found there.

The hilly foreland of the *Chapada* currently provides retreats for the smallholders in areas which have emanated from state-run colonisation projects or illegal occupations. Here, smallholder families can only survive because of the relative proximity of an urban settlement and the associated employment opportunities.

The excellent agricultural suitability of the soils in the central, less hilly area of the region provided an adequate economic basis for the better-off smallholders from the private colonisation projects. However, the first signs of degradation threatening the existence of the settlers can be observed here as well.

A spatial pattern of criticality arises from the combination of these socio-economic and ecological factors and processes. Its foundations were laid in the initial phase of pioneer frontier development: A ring of areas of high criticality, where the response systems to prevent the collapse of current socio-economic structures are dysfunctional, surrounds the central areas which have low levels of criticality. In the central areas, however, there is a latent risk, as the smallholders have only limited strategies to overcome crises.

4. Smallholders in Baixo Alegre between ecological degradation and securing survival

The settlement of Baixo Alegre is situated to the north of the municipality of Rio Branco and is one of the most endangered regions with a high level of criticality. It serves as an example to illustrate the displacement processes to which smallholders at pioneer frontiers are subject, as well as the factors contributing to their vulnerability, and their survival strategies.

Directly adjoining the state-run colonisation project of Rio Branco, the *comunidade* Baixo Alegre emerged when an extensively operated cattle farm was occupied at the beginning of the 1980s. In particular smallholders who had already been displaced from other colonisation projects in this region, found a retreat on this land. However, the settlers could barely earn their livelihood cultivating parcels of land only a few hectares in extent, in hilly terrain entirely unsuitable for agriculture. Only a few years after clearing the forest, the cultivation of staple foods led to soil erosion processes which could only be stopped by sowing grass for pasture. This forced many families to migrate, as the parcels of land were too small for cattle farming, and many smallholders did not have the funds to buy dairy cattle. Moreover, the attraction of the pioneer frontiers which were booming in Rondônia and northern Mato-Grosso during these years, was so overwhelming that more than half of the original settler families sought a better future there.

This migration led to a new, spatially differentiated pattern of land ownership (see Fig. 4). Migrants sold their abandoned plots in the ecologically unfavourable, inaccessible slope locations to neighbouring farmers, also fighting for their own survival, or, as land was almost totally worthless, to trade them for small amounts of food. Nevertheless, the parcels of land remained small.

In the moderately hilly valley of the Córrego do Pito the plots were sold for higher prices, so that only families with sufficient capital could purchase them. Usually these settlers had already successfully cultivated land in adjoining areas and wanted to enlarge their farms, or they were city dwellers building up an economic stand-by. Even adjacent large-scale farms bought into the *comunidade* little by little. Thus, the economically better-structured farms managed to occupy the favourable areas in the valley, while the destitute smallholders were displaced into slope locations. As a result, the entirely marginalised families were forced to move into far-off survival spaces, i.e. into new pioneer frontier areas.

These ecological factors of smallholder vulnerability coincided with a series of regional development processes. With increased soil erosion and the decline in the productivity of the land and in staple food production, the corresponding marketing structures collapsed

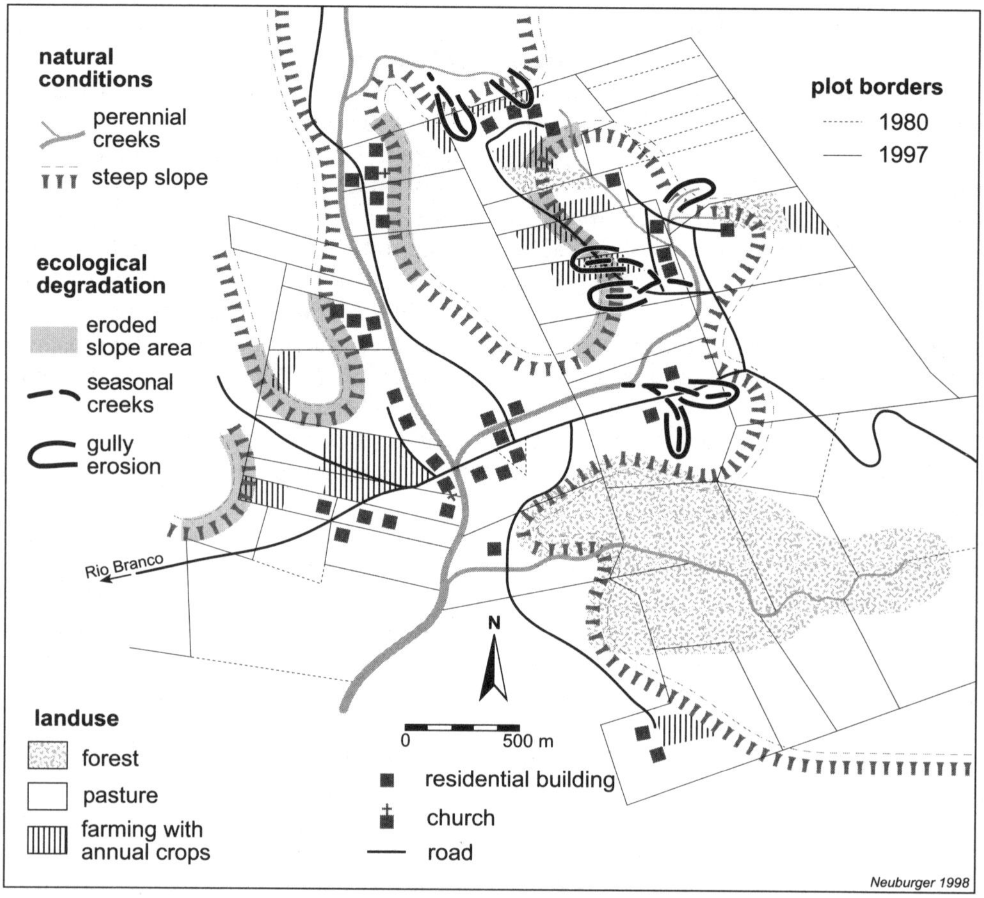

Fig. 4: Ecological degradation in Baixo Alegre

as well. Thus production for sale on the local markets made no sense for the individual any more.

The shift to dairy cattle farming, which could only be considered by those able to expand their acreage, resulted in only short-term relief during the economic crisis as milk prices have also been falling since the 1990s. To this day, growing economic pressure forces the smallholders to use their parcels intensively in ecologically fragile areas. Planting densities which are too high and the extension of agriculture to steep slope locations increase ecological degradation and the associated vulnerability of the smallholder families (see Fig. 4).

The economic crisis in the entire region also affects local administration. Because of their close connection with the local upper class, the political elite rarely use the limited financial resources for the good of the marginalised population, and as a consequence, the infrastructure in rural areas is increasingly dilapidated. The basic infrastructure, which was built as a part of the Polonoroeste Program in the 1980s, is no longer maintained. Therefore the school in Baixo Alegre was reduced to only a few classes. Yet, both parents and children are reluctant to attend city-schools, and thus they remain with virtually no school education. The settlers avoid using health services in the city, if possible. These are totally congested because the administration does not expand them in accordance with the rapid growth of the city's population. Even roads destroyed by heavy rains are only sporadically repaired, usually during election campaigns. As a result, the smallholders from Baixo Alegre are not able to sell their fresh milk and are thus must make cheese for sale on the local market. The interweaving of political-institutional factors contributing to smallholder vulnerability with economic and ecological processes is particularly apparent here.

Finally, the social component of vulnerability is also closely related to the above-mentioned processes. The few smallholder families remaining in Baixo Alegre see their social networks collapse because of the heavy migration out of the *comunidade*; networks which play a central role in coping with everyday life and also in overcoming crises. Mutual help and co-operative work are no longer possible, and thus the individual families are on their own. In many cases, recourse to social networks (mostly relatives) in the urban areas is the only way to cope with emergency situations. Migration into the cities is a likely result.

Only a few smallholder families can withstand the pressure towards displacement. Their multi-layered vulnerability produces a multitude of survival strategies, which the Baixo Alegre smallholders combine in different ways. The typical and most important coping strategy of "losers out" in the frontier development is migration, either into the city or to new pioneer frontiers. During the early phase of migration, usually entire families gave up their farms. Within the last few years, individual family members, for the most part from the younger generations, migrated into the cities, in many cases leaving behind only core families. This development is reflected in the number of children in the *comunidade,* which has fallen from about 300 in 1986 to only 30 in 1997. This survival strategy, with the associated collapse of social networks, leads to increased vulnerability.

Another strategy which smallholders employ in dealing with crises, is the reduction of their household expenses. To this end subsistence production is increased. This refers not only to the cultivation of staple food, but also includes the production of simple equipment and tools from local materials. Yet it is this very strategy which often contributes to the overuse of local natural resources. Thus, for example, intensive agriculture causes serious erosion damage, further lowering productivity. The rate of tree-felling exceeds the ability of the remaining secondary forest to regenerate. Hence the smallholder families' vulnerability is increased yet again.

In order to earn the money necessary for medication, clothes and school equipment, the families tap various sources of income ranging from the refinement and sale of farm products to accepting occasional jobs on neighbouring large-scale farms or in the cities. Some older family members also receive a state pension, and thus small sums of money are available fairly regularly. Although this diversification of income lowers the families' immediate susceptibility to risks, their incorporation into regional business cycles based on their exploitation makes them vulnerable to economic crises which they themselves cannot influence.

In order to counteract this danger and put agricultural production back into the foreground, many smallholders joined the local farmers' association. But after a few years the better-structured farmers from the *comunidade* exploited the association to their own advantage.

Today, all that is left to the Baixo Alegre smallholders is the hope placed in the political vigour of the landless workers' movement MST (*Movimento dos Trabalhadores Rurais sem Terra*). It became active in the state of Mato Grosso during the last few years and managed to organise a couple of spectacular land occupations (for an introduction to MST see Fernandes 1996). However, both the economic elite (especially owners of large estates) and political representatives are working against the interests of MST and are attempting to impede the success of their operations.

5. Prospects for the future

In the preceding analysis of three aspects of the Brazilian pioneer frontier, the correlation of spatial structures, sectoral developments and historical processes has become evident. The examination of different levels has shown that the development of the Brazilian pioneer frontiers is incorporated into global and national power and market structures. The utilisation of peripheral areas as retreats for marginalised groups, smallholders in particular, is reflected down to the local level. Furthermore, national policies, especially agrarian policy following the Brazilian development model, determine the viability of peasant farms. Recently Brazil's high foreign debt as an expression of world-economic structures, the globally established development aim of sustainability, and the national financial crisis have influenced the development of the Amazon region down to the local level.

The analysis of different social realms reveals the dominance of disparate structures at the various levels, as well as the personal and structural interconnections of the economic and political elite. It is virtually impossible for marginalised groups to articulate and enforce their interests within these structures. Thus they are often displaced into ecologically fragile locations, thereby increasing their vulnerability. Due to a variety of constraints, they can only counteract displacement processes with the aid of survival strategies which are for the most part unlikely to reduce their risk susceptibility. In the last few years, growing processes of globalisation have strengthened these groups somewhat, especially as international NGOs take up their cause.

The historical analysis demonstrates the reproduction of political-economic power and actor constellations over time. The disparate structures which determine socio-economic, political and ecological processes at the global, national and regional/local level to this day, were created during the colonial period. The one-sided relations of dependency between the colonial power and the colony are continued today in the imbalance between developed and undeveloped countries within the global economic system. In a national context, the utilisation of the Amazon region as the "last" pioneer frontier is probably the end of frontier expansion in Brazil. The smallholder groups who have been exposed to marginalisation and displacement processes ever since the colonial period, which are mirrored in socio-economic structures at the local level, must be considered in their historical context.

Against this background, ecological degradation processes in the Amazon pioneer frontier regions can be understood as a product of global, national and regional/local structures, within which particularly vulnerable groups (here smallholders) have been displaced into ecologically fragile areas in the historical process. The exploitation of the natural resources of this area of retreat as a coping strategy is part of their logic of action, balancing between vulnerability and the constraints of the respective everyday situation. The most recent tendencies of globalisation partially break these dependency structures and offer new chances for marginalised groups. Functional response systems, which can offer the prospect of survival and a sustainable economic and social perspective even to the weakest members of society, naturally presuppose political will to allow for sustainable development on the part of the national elite as well as on the part of the regional/local elite.

References

Blaikie, P. and Brookfield, H. (eds.) (1987): Land degradation and society. London, New York.

Blaikie, P., Cannon, J. D. and Wisner, B. (1994): At risk. Natural hazards, people's vulnerability and disasters. London, New York.

Bohle, H.-G. (1998): Strategien der Überlebenssicherung und Verwundbarkeit in Entwicklungsländern. – Rundbrief Geographie 149, pp. 13-16.

Bryant, R. L. and Bailey, S. (1997): Third World political ecology. London, New York.

Coy, M. and Lücker, R. (1993): Der brasilianische Mittelwesten. Wirtschafts- und sozialgeographischer Wandel eines peripheren Agrarraumes. – Tübinger Beiträge zur Geographischen Lateinamerika-Forschung 9. Tübingen.

Coy, M. and Neuburger, M. (1999): As frentes pioneiras na Amazônia brasileira entre globalização e sustentabilidade. CeHu (Centro de Estudios Alexander von Humboldt) (ed.): Primer Encuentro Internacional Humboldt; Buenos Aires (CD-Rom).

Dean, W. (1996): A ferro e fogo. A história e a devastação da mata atlântica brasileira. São Paulo.

Fernandes, B. Mançano (1996): MST – Formação e territorialização em São Paulo. São Paulo.

Geist, H. (1992): Die orthodoxe und politisch-ökologische Sichtweise von Umweltdegradierung. – Die Erde 123, pp. 283–295.

Goudie, A. (1994): Mensch und Umwelt. Eine Einführung. Heidelberg, Berlin, Oxford.

Guidon, N. (1992): As ocupações pré-históricas do Brasil (excetuando a Amazônia). M. Carneiro da Cunha (ed.): História dos índios no Brasil. São Paulo. pp. 37–52.

Hall, A. (1997): Sustaining Amazonia. Grassroots action for productive conservation. Manchester, New York.

Hewitt, K. (1997): Regions of risk. A geographical introduction to disasters. Herlow.

Holling, C. S. (ed.) (1980): Adaptive environmental assessment and management. – International Series on Applied Systems Analysis 3. Chichester, New York.

Kasperson, R. E., Kasperson, J. X., Turner II, B. L., Dow, K. and Meyer, W. B. (1995): Critical environmental regions. Concepts, distinctions, and issues. J. X. Kasperson, R. E. Kasperson and B. L. Turner II (eds.): Regions at risk. Comparisons of threatened environments. Tokio, New York, Paris. pp. 1–41.

Kohlhepp, G. (1979): Brasiliens problematische Antithese zur Agrarreform: Agrarkolonisation in Amazonien. Evaluierung wirtschafts- und sozialgeographischer Prozeßabläufe an der Peripherie im Lichte wechselnder agrarpolitischer Strategien. H. Elsenhans (ed.): Agrarreform in der Dritten Welt. Frankfurt am Main, New York. pp. 471–504.

Kohlhepp, G. (1987): Amazonien. – Problemräume der Welt 8. Köln.

Kohlhepp, G. (1990): Landnutzungs-Sukzessionen im nördlichen Paraná (Südbrasilien) am Beispiel regionaler und betrieblicher Strukturwandlungen. B. Mohr, K. Sonntag and J. Stadelbauer (eds.): Räumliche Strukturen im Wandel. Teil B: Beiträge zur Agrarwirtschaft der Tropen. – Freiburger Geographische Hefte 30. Freiburg. pp. 45–68.

Lücker, R. (1986): Agrarräumliche Entwicklungsprozesse im Alto-Uruguai-Gebiet (Südbrasilien). – Tübinger Beiträge zur Geographischen Lateinamerika-Forschung 2. Tübingen.

Manshard, W. and Mäckel, R. (1995): Umwelt und Entwicklung in den Tropen. Naturpotential und Landnutzung. Darmstadt.

Martins, J. de Souza (1990): Os camponeses e a política no Brasil. As lutas sociais no campo e seu lugar no processo político. 4th ed. Petrópolis.

Martins, J. de Souza (1997): Fronteira. A degradação do outro nos confins do humano. São Paulo.

Painter, M. and Durham, W. H. (eds.) (1995): The social causes of environmental destruction in Latin America. Ann Arbor.

Peet, R. and Watts, M. (eds.) (1996): Liberation ecologies. Environment, development, social movements. London, New York.

Remppis, M. (1998): Chancen und Risiken eines liberalisierten Weltagrarmarkts. Perspektiven der Rindfleischproduktion in einem peripheren Raum Brasiliens am Beispiel von Mato Grosso. G. Kohlhepp. And M. Coy (eds.): Mensch-Umwelt-Beziehungen und nachhaltige Entwicklung in der Dritten Welt. – Tübinger Beiträge zur Geographischen Lateinamerikaforschung 119. Tübingen. pp. 87–107.

Roosevelt, A. Curtenius (1992): Arqueologia amazônica. M. Carneiro da Cunha (ed.): História dos índios no Brasil. São Paulo, pp. 53–86.

Schmink, M. and Wood, Ch. H. (1986): The political ecology of Amazonia. P. D. Little and M. M. Horowitz (eds.): Lands at risk in the Third World. Local-level perspectives. Boulder, London. pp. 38–57.

Schmink, M. and Wood, Ch. H. (1992): Contested frontiers in Amazonia. New York.

Woortmann, E.F. and Woortmann, K. (1997): O trabalho da terra. A lógica e a simbólica da lavoura camponesa. Brasília.

Address of the author: Dr. Martina Neuburger, Geographisches Institut, Universität Tübingen, Hölderlinstr. 12, D-72074 Tübingen
Martina.neuburger@uni-tuebingen.de

First published in Geographische Zeitschrift vol. 88/ 2000 (1), pp. 21–35